T0269086

Symmetry Methods for Differential Equations

Symmetry is the key to solving differential equations. There are many well-known techniques for obtaining exact solutions, but most of them are merely special cases of a few powerful symmetry methods. These methods can be applied to differential equations of an unfamiliar type; they do not rely on special "tricks." Instead, a given differential equation can be made to reveal its symmetries, which are then used to construct exact solutions.

This book is a straightforward introduction to symmetry methods; it is aimed at applied mathematicians, physicists, and engineers. The presentation is informal, with many worked examples. It is written at a level suitable for postgraduates and advanced undergraduates. The reader should be able to master the main techniques quickly and easily.

This text contains several new methods that will interest those whose research involves symmetries. In particular, methods for obtaining discrete symmetries and first integrals are described.

Peter Hydon is a Lecturer in Mathematics at the University of Surrey.

Cambridge Texts in Applied Mathematics

Maximum and Minimum Principles
M.J. Sewell

Solitons
P.G. Drazin and R.S. Johnson

The Kinematics of Mixing
J.M. Ottino

Introduction to Numerical Linear Algebra and Optimisation
Philippe G. Ciarlet

Integral Equations
David Porter and David S.G. Stirling

Perturbation Methods
E.J. Hinch

The Thermomechanics of Plasticity and Fracture
Gerard A. Maugin

Boundary Integral and Singularity Methods for Linearized Viscous Flow
C. Pozrikidis

Nonlinear Wave Processes in Acoustics
K. Naugolnykh and L. Ostrovsky

Nonlinear Systems
P.G. Drazin

Stability, Instability and Chaos
Paul Glendinning

Applied Analysis of the Navier–Stokes Equations
C.R. Doering and J.D. Gibbon

Viscous Flow
H. Ockendon and J.R. Ockendon

Scaling, Self-Similarity and Intermediate Asymptotics
G.I. Barenblatt

A First Course in the Numerical Analysis of Differential Equations
A. Iserles

Complex Variables: Introduction and Applications
M.J. Ablowitz and A.S. Fokas

Mathematical Models in the Applied Sciences
A.C. Fowler

Thinking About Ordinary Differential Equations
R. O'Malley

A Modern Introduction to the Mathematical Theory of Water Waves
R.S. Johnson

The Space–Time Ray Method
V.M. Babich, I. Molotkov and V.S. Buldyrev

Rarefied Gas Dynamics
Carlo Cercignani

Symmetry Methods for Differential Equations
Peter E. Hydon

High Speed Flow
C.J. Chapman

Symmetry Methods for Differential Equations
A Beginner's Guide

PETER E. HYDON
Department of Mathematics & Statistics
University of Surrey

CAMBRIDGE UNIVERSITY PRESS
Cambridge, New York, Melbourne, Madrid, Cape Town, Singapore, São Paulo

Cambridge University Press
The Edinburgh Building, Cambridge CB2 2RU, UK

Published in the United States of America by Cambridge University Press, New York

www.cambridge.org
Information on this title: www.cambridge.org/9780521497039

First published 2000

A catalogue record for this publication is available from the British Library

Library of Congress Cataloguing in Publication data

Hydon, Peter E. (Peter Ellsworth), 1960–
Symmetry methods for differential equations : a beginner's guide /
Peter E. Hydon.
p. cm. – (Cambridge texts in applied mathematics)
Includes bibliographical references and index.
ISBN 0-521-49703-5
1. Differential equations–Numerical solutions. 2. Symmetry.
3. Mathematical physics. I. Title. II. Series.
QC20.7.D5H93 2000
530.15′535 – dc21 99-31354
CIP

ISBN-13 978-0-521-49703-9 hardback
ISBN-10 0-521-49703-5 hardback

ISBN-13 978-0-521-49786-2 paperback
ISBN-10 0-521-49786-8 paperback

Transferred to digital printing 2005

To
Alison
Christopher
Rachel
and
Katy
who waited patiently for me
to come out of my study.
The wait is over.

Contents

Preface

There are many ingenious techniques for obtaining exact solutions of differential equations, but most work only for a very limited class of problems. How can one solve differential equations of an unfamiliar type?

Surprisingly, most well-known techniques have a common feature: they exploit *symmetries* of differential equations. It is often quite easy to find symmetries of a given differential equation (even an unfamiliar one) and to use them systematically to obtain exact solutions. Symmetries can also be used to simplify problems and to understand bifurcations of nonlinear systems.

More than a century ago, the Norwegian mathematician Sophus Lie put forward many of the fundamental ideas behind symmetry methods. Most of these ideas are essentially simple, but are so far reaching that they are still the basis of much research. As an applied mathematician, I have found symmetry methods to be invaluable. They are fairly easy to master and provide the user with a powerful range of tools for studying new equations. I believe that no one who works with differential equations can afford to be ignorant of these methods.

This book introduces applied mathematicians, engineers, and physicists to the most useful symmetry methods. It is aimed primarily at postgraduates and those involved in research, but there is sufficient elementary material for a one-semester undergraduate course. (Over the past five years, I have taught these methods to both undergraduates and postgraduates.) Bearing in mind the interests and needs of the intended readership, the book focuses on techniques. These are described and justified informally, without a "theorem–proof" format. I have tried to present the theory straightforwardly, sacrificing rigour and generality (where necessary) in order to communicate the most useful results clearly.

The topics are arranged so as to provide a graded introduction. Thus the reader can see symmetry methods applied at an early stage, without first having to absorb much new notation. As the book progresses, the methods are generalized and extended. Practice is essential to develop skill in using symmetry methods; readers are urged to try the exercises at the end of each chapter. Solutions and hints for some exercises are available at the end of the book.

Here are some suggestions for those wishing to use this book as the basis of a lecture course. The first six chapters consist of core material on ordinary differential equations. In my experience, this is sufficient for a one-semester undergraduate course. For a postgraduate course, Chapters 8 and 9 (which deal with basic symmetry methods for partial differential equations) should also be included. I strongly recommend that students learn how to use an appropriate computer algebra package, because symmetry calculations can be lengthy (particularly for partial differential equations). I have briefly outlined some packages that are currently available at no cost to the user.

The remaining chapters outline some recent developments. These are selected on the grounds that they are widely applicable and easy to master. Some of these topics have not previously been described at an elementary level. I have omitted several techniques on the grounds that they are difficult to describe accurately without using complicated mathematical ideas. My aim throughout has been to enable the reader to become proficient in the most useful symmetry methods.

Peter E. Hydon
January 1999

Acknowledgements

I thank those who have read all or parts of the manuscript and have suggested ways to improve it: David Gammack, Nick Hill, Fiona Laine-Pearson, Tassos Makris, Liz Mansfield, Sebastian Reich, and Sue Todd. I also thank Alan Harvey, David Tranah, Linda and Peter Clist, and my family for their unfailing encouragement and help. I am indebted to Peter Clarkson, whose enthusiasm for symmetry methods is infectious!

1

Introduction to Symmetries

I know it when I see it.

(Justice Potter Stewart: Jacoblellis v. Ohio, 378 U.S. 184, 197 [1964])

1.1 Symmetries of Planar Objects

In order to understand symmetries of differential equations, it is helpful to consider symmetries of simpler objects. Roughly speaking, a symmetry of a geometrical object is a transformation whose action leaves the object apparently unchanged. For instance, consider the result of rotating an equilateral triangle anticlockwise about its centre. After a rotation of $2\pi/3$, the triangle looks the same as it did before the rotation, so this transformation is a symmetry. Rotations of $4\pi/3$ and 2π are also symmetries of the equilateral triangle. In fact, rotating by 2π is equivalent to doing nothing, because each point is mapped to itself. The transformation mapping each point to itself is a symmetry of any geometrical object: it is called the *trivial symmetry*.

Symmetries are commonly used to classify geometrical objects. Suppose that the three triangles illustrated in Fig. 1.1 are made from some rigid material, with indistinguishable sides. The symmetries of these triangles are readily found by experiment. The equilateral triangle has the trivial symmetry, the rotations described above, and flips about the three axes marked in Fig. 1.1(a). These flips are equivalent to reflections in the axes. So an equilateral triangle has six distinct symmetries. The isoceles triangle in Fig. 1.1(b) has two: a flip (as shown) and the trivial symmetry. Finally, the triangle with three unequal sides in Fig. 1.1(c) has only the trivial symmetry.

There are certain constraints on symmetries of geometrical objects. Each symmetry has a unique inverse, which is itself a symmetry. The combined action of the symmetry and its inverse upon the object (in either order) leaves

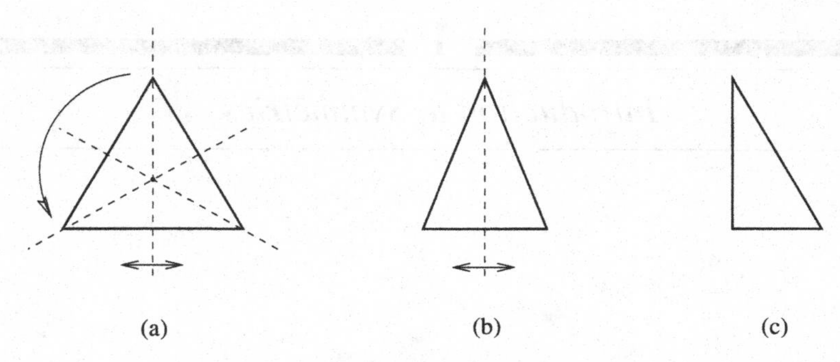

Fig. 1.1. Some triangles and their symmetries.

the object unchanged. For example, let Γ denote a rotation of the equilateral triangle by $2\pi/3$. Then Γ^{-1} (the inverse of Γ) is a rotation by $4\pi/3$.

For simplicity, we restrict attention to symmetries that are *smooth*. (This somewhat technical requirement is not greatly restrictive, and it frees us from the need to consider pathological examples.) If x denotes the position of a general point of the object, and if

$$\Gamma : x \mapsto \hat{x}(x)$$

is any symmetry, then we assume that $\hat{x}$ is infinitely differentiable with respect to x. Moreover, since Γ^{-1} is also a symmetry, x is infinitely differentiable with respect to $\hat{x}$. Thus Γ is a (C^∞) *diffeomorphism*, that is, a smooth invertible mapping whose inverse is also smooth.

Symmetries are also required to be *structure preserving*. It is usual for geometrical objects to have some structure which (loosely speaking) describes what the object is made from. To use an analogy from continuum mechanics, the structure is the constitutive relation for the object. Earlier, we considered symmetries of triangles made from a rigid material. The only transformations under which a triangle remains rigid are those which preserve the distance between any two points on the triangle, namely translations, rotations, and reflections (flips). These transformations are the only possible symmetries, because all other transformations fail to preserve the rigid structure. However, if the triangles are made from an elastic material such as rubber, the class of structure-preserving transformations is larger, and new symmetries may be found. For example, a triangle with three unequal sides can be stretched into an equilateral triangle, then rotated by $2\pi/3$ about its centre, and finally stretched so as to appear to have its original shape. This transformation is not a symmetry of a rigid triangle. Clearly, the structure associated with a geometrical object has a considerable influence upon the set of symmetries of the object.

In summary, a transformation is a symmetry if it satisfies the following:

(S1) The transformation preserves the structure.
(S2) The transformation is a diffeomorphism.
(S3) The transformation maps the object to itself [e.g., a planar object in the (x, y) plane and its image in the $(\hat{x}, \hat{y})$ plane are indistinguishable].

Henceforth, we restrict attention to transformations satisfying (S1) and (S2). Such transformations are symmetries if they also satisfy (S3), which is called the *symmetry condition.*

A rigid triangle has a finite set of symmetries. Many objects have an infinite set of symmetries. For example, the (rigid) unit circle

$$x^2 + y^2 = 1$$

has a symmetry

$$\Gamma_\varepsilon : (x, y) \mapsto (\hat{x}, \hat{y}) = (x \cos\varepsilon - y \sin\varepsilon,\ x \sin\varepsilon + y \cos\varepsilon)$$

for each $\varepsilon \in (-\pi, \pi]$. In terms of polar coordinates,

$$\Gamma_\varepsilon : (\cos\theta, \sin\theta) \mapsto \big(\cos(\theta + \varepsilon), \sin(\theta + \varepsilon)\big),$$

as shown in Fig. 1.2. Hence the transformation is a rotation by ε about the centre of the circle. It preserves the structure (rotations are rigid), and it is smooth and invertible (the inverse of a rotation by ε is a rotation by $-\varepsilon$). To prove that the symmetry condition (S3) is satisfied, note that

$$\hat{x}^2 + \hat{y}^2 = x^2 + y^2,$$

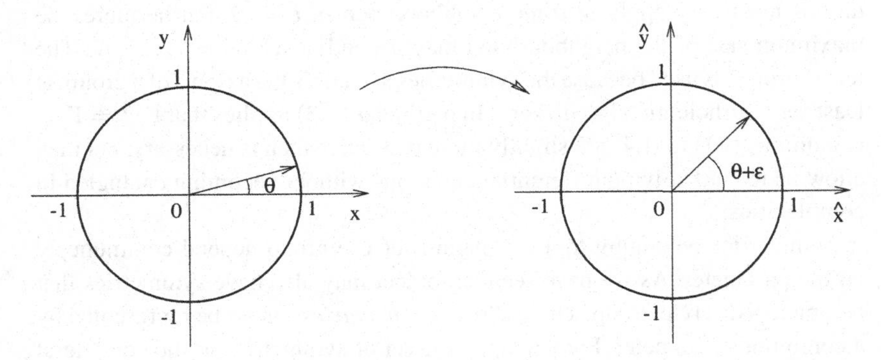

Fig. 1.2. Rotation of the unit circle.

and therefore

$$\hat{x}^2 + \hat{y}^2 = 1 \quad \text{when} \quad x^2 + y^2 = 1.$$

The unit circle has other symmetries, namely reflections in each straight line passing through the centre. It is not difficult to show that every reflection is equivalent to the reflection

$$\Gamma_R : (x, y) \mapsto (-x, y)$$

followed by a rotation Γ_ε.

The infinite set of symmetries Γ_ε is an example of a *one-parameter Lie group*. This class of symmetries is immensely useful and is the key to constructing exact solutions of many differential equations. Suppose that an object occupying a subset of $\mathbb{R}^N$ has an infinite set of symmetries

$$\Gamma_\varepsilon : x^s \mapsto \hat{x}^s(x^1, \ldots, x^N; \varepsilon), \qquad s = 1, \ldots, N,$$

where ε is a real parameter, and that the following conditions are satisfied.

(L1) Γ_0 is the trivial symmetry, so that $\hat{x}^s = x^s$ when $\varepsilon = 0$.
(L2) Γ_ε is a symmetry for every ε in some neighbourhood of zero.
(L3) $\Gamma_\delta \Gamma_\varepsilon = \Gamma_{\delta+\varepsilon}$ for every δ, ε sufficiently close to zero.
(L4) Each $\hat{x}^s$ may be represented as a Taylor series in ε (in some neighbourhood of $\varepsilon = 0$), and therefore

$$\hat{x}^s(x^1, \ldots, x^N; \varepsilon) = x^s + \varepsilon \xi^s(x^1, \ldots, x^N) + O(\varepsilon^2), \qquad s = 1, \ldots, N.$$

Then the set of symmetries Γ_ε is a one-parameter local Lie group. The term "local" (which we shall usually omit hereafter) refers to the fact that the conditions need only apply in some neighbourhood of $\varepsilon = 0$. Furthermore, the maximum size of the neighbourhood may depend on x^s, $s = 1, \ldots, N$. The term "group" is used because the symmetries Γ_ε satisfy the axioms of a group, at least for ε sufficiently close to zero. In particular, (L3) implies that $\Gamma_\varepsilon^{-1} = \Gamma_{-\varepsilon}$. Conditions (L1) to (L4) are slightly more restrictive than is necessary, but they allow us to start solving differential equations without becoming entangled in complexities.

Symmetries belonging to a one-parameter Lie group depend continuously on the parameter. As we have seen, an object may also have symmetries that belong to a discrete group. These *discrete symmetries* cannot be represented by a continuous parameter. For example, the set of symmetries of the equilateral triangle has the structure of the dihedral group D_3, whereas the two symmetries of the isoceles triangle form the cyclic group $\mathbb{Z}_2$. Discrete symmetries are useful in many ways, as described at the end of the book. Until then, we shall focus on parametrized Lie groups of symmetries, which are easier to find and use. For brevity, we refer to such symmetries as *Lie symmetries*.

For most of the time, we shall study the functions $\hat{x}^s$ directly, without reference to any ideas from group theory. Therefore it is convenient to simplify the notation by abbreviating

$$\Gamma_\varepsilon : (x^1, \dots, x^N) \mapsto (\hat{x}^1, \dots, \hat{x}^N) = \cdots$$

to

$$(\hat{x}^1, \dots, \hat{x}^N) = \cdots .$$

Suffix notation is useful for stating general results, but we shall avoid using it in examples, as far as possible. Variables will be named $x, y, \dots$ in preference to $x^1, x^2, \dots$.

1.2 Symmetries of the Simplest ODE

What are the symmetries of ordinary differential equations (ODEs)? To begin to answer this question, consider the simplest ODE of all, namely

$$\frac{dy}{dx} = 0. \tag{1.1}$$

The set of all solutions of the ODE is the set of lines

$$y(x) = c, \qquad c \in R,$$

which fills the (x, y) plane. The ODE (1.1) is represented geometrically by the set of all solutions, and so any symmetry of the ODE must necessarily map the solution set to itself. More formally, the symmetry condition (S3) requires that the set of solution curves in the (x, y) plane must be indistinguishable from its image in the $(\hat{x}, \hat{y})$ plane, and therefore

$$\frac{d\hat{y}}{d\hat{x}} = 0 \quad \text{when} \quad \frac{dy}{dx} = 0. \tag{1.2}$$

A smooth transformation of the plane is invertible if its Jacobian is nonzero, so we impose the further condition

$$\hat{x}_x \hat{y}_y - \hat{x}_y \hat{y}_x \neq 0. \tag{1.3}$$

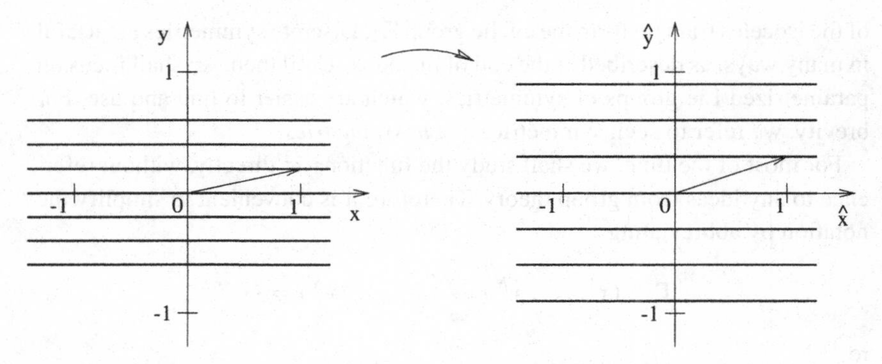

Fig. 1.3. Solutions of (1.1) transformed by a scaling, (1.5).

(Throughout the book, variable subscripts denote partial derivatives, e.g., $\hat{x}_x$ denotes $\frac{\partial \hat{x}}{\partial x}$.) A particular solution curve will be mapped to a (possibly different) solution curve, and so

$$\hat{y}(x, c) = \hat{c}(c), \qquad \forall c \in \mathbb{R}. \tag{1.4}$$

Here x is regarded as a function of $\hat{x}$ and c that is obtained by inverting

$$\hat{x} = \hat{x}(x, c).$$

The ODE (1.1) has many symmetries, some of which are obvious from Fig. 1.3. There are discrete symmetries, such as reflections in the x and y axes. Lie symmetries include scalings of the form

$$(\hat{x}, \hat{y}) = (x, e^{\varepsilon} y), \qquad \varepsilon \in \mathbb{R}. \tag{1.5}$$

[Figure 1.3 depicts the effect of the scalings (1.5) on only a few solution curves; if all solution curves could be shown, the two halves of the figure would be identical.] Every translation,

$$(\hat{x}, \hat{y}) = (x + \varepsilon_1, y + \varepsilon_2), \qquad \varepsilon_1, \varepsilon_2 \in \mathbb{R}, \tag{1.6}$$

is a symmetry. The set of all translations depends upon two parameters, ε_1 and ε_2. By setting ε_1 to zero, we obtain the one-parameter Lie group of translations in the y direction. Similarly, the one-parameter Lie group of translations in the x direction is obtained by setting ε_2 to zero. The set of translations (1.6) is a *two*-parameter Lie group, which can be regarded as a composition of the one-parameter Lie groups of translations parametrized by ε_1 and ε_2 respectively. Roughly speaking, symmetries belonging to an R-parameter Lie group can be regarded as a composition of symmetries from R one-parameter Lie groups.

Not every one-parameter Lie group is useful. For example, a translation (1.6) maps a solution curve $y = c$ to the curve $\hat{y} = c + \varepsilon_2$. If $\varepsilon_2 = 0$, any solution curve is mapped to itself by the symmetry. This is obvious, because translations in the x direction move points along the curves of constant y. Symmetries that map every solution curve to itself are described as *trivial*, even if they move points along the curves.

The ODE (1.1) is extremely simple, and so all of its symmetries can be found. Differentiating (1.4) with respect to x, we obtain

$$\hat{y}_x(x, c) = 0, \qquad \forall c \in \mathbb{R}.$$

Therefore, taking (1.3) into account, the symmetries of (1.1) are of the form

$$(\hat{x}, \hat{y}) = (f(x, y), g(y)), \qquad f_x \neq 0, \quad g_y \neq 0, \tag{1.7}$$

where f and g are assumed to be smooth functions of their arguments. The ODE has a very large family of symmetries. (Perhaps surprisingly, so does every first-order ODE.)

We were able to use the known general solution of (1.1) to derive (1.2), which led to the result (1.7). However, we could also have obtained this result directly from (1.2), as follows. On the solution curves, y is a function of x, and hence $\hat{x}(x, y)$ and $\hat{y}(x, y)$ may be regarded as functions of x. Then, by the chain rule, (1.2) can be rewritten as

$$\frac{d\hat{y}}{d\hat{x}} = \frac{D_x\hat{y}}{D_x\hat{x}} = 0 \qquad \text{when} \quad \frac{dy}{dx} = 0,$$

where D_x denotes the *total derivative* with respect to x:

$$D_x = \partial_x + y'\partial_y + y''\partial_{y'} + \cdots. \tag{1.8}$$

(The following notation is used throughout the book: ∂_x denotes $\frac{\partial}{\partial x}$, etc; y' denotes $\frac{dy}{dx}$, etc.) Therefore (1.2) amounts to

$$\frac{\hat{y}_x + y'\hat{y}_y}{\hat{x}_x + y'\hat{x}_y} = 0 \qquad \text{when} \quad y' = 0,$$

that is,

$$\frac{\hat{y}_x}{\hat{x}_x} = 0.$$

Hence (1.7) holds. The advantage of using the symmetry condition in the form (1.2) is that one can obtain information about the symmetries without having to know the solution of the differential equation in advance. This observation is fundamental, for it suggests that it might be possible to find symmetries of a given differential equation whose solution is unknown.

1.3 The Symmetry Condition for First-Order ODEs

The symmetries of $y' = 0$ are easily visualized, because the solution curves are parallel lines. It may not be possible to find symmetries of a complicated first-order ODE by looking at a picture of its solution curves. Nevertheless, the symmetry condition requires that any symmetry maps the set of solution curves in the (x, y) plane to an identical set of curves in the $(\hat{x}, \hat{y})$ plane. Consider a first-order ODE,

$$\frac{dy}{dx} = \omega(x, y). \tag{1.9}$$

(For simplicity, we shall restrict attention to regions of the plane in which ω is a smooth function of its arguments.) The symmetry condition for (1.9) is

$$\frac{d\hat{y}}{d\hat{x}} = \omega(\hat{x}, \hat{y}) \qquad \text{when} \qquad \frac{dy}{dx} = \omega(x, y). \tag{1.10}$$

As before, we regard y as a function of x (and a constant of integration) on the solution curves. Then (1.10) yields

$$\frac{D_x \hat{y}}{D_x \hat{x}} = \frac{\hat{y}_x + y' \hat{y}_y}{\hat{x}_x + y' \hat{x}_y} = \omega(\hat{x}, \hat{y}) \qquad \text{when} \qquad \frac{dy}{dx} = \omega(x, y).$$

Therefore the symmetry condition for the first-order ODE (1.9) is equivalent to the constraint

$$\frac{\hat{y}_x + \omega(x, y) \hat{y}_y}{\hat{x}_x + \omega(x, y) \hat{x}_y} = \omega(\hat{x}, \hat{y}), \tag{1.11}$$

together with the requirement that the mapping should be a diffeomorphism. It may be possible to determine some or all of the symmetries of a given ODE from (1.11). One approach is use an *ansatz*, that is, to look for a symmetry of a particular form.

Example 1.1 Consider the ODE

$$\frac{dy}{dx} = y. \tag{1.12}$$

The constraint (1.11) implies that every symmetry of (1.12) satisfies the partial differential equation (PDE)

$$\frac{\hat{y}_x + y\hat{y}_y}{\hat{x}_x + y\hat{x}_y} = \hat{y}.$$

Rather than trying to find the general solution of this PDE, let us see whether or not there are symmetries that satisfy a simple ansatz. For example, are there any symmetries mapping y to itself? If so, then

$$(\hat{x}, \hat{y}) = (\hat{x}(x, y), y),$$

and the constraint (1.11) reduces to

$$\frac{y}{\hat{x}_x + y\hat{x}_y} = y.$$

Therefore (taking (1.3) into account),

$$\hat{x}_x + y\hat{x}_y = 1, \qquad \hat{x}_x \neq 0.$$

There are many symmetries of this type; the simplest are the Lie symmetries

$$(\hat{x}, \hat{y}) = (x + \varepsilon, y), \qquad \varepsilon \in \mathbb{R}. \tag{1.13}$$

Earlier, we found that translations in the x direction are trivial symmetries of $y' = 0$; are they also trivial symmetries of (1.12)? The general solution of (1.12) is easily found; it is

$$y = c_1 e^x, \qquad c_1 \in \mathbb{R}.$$

A translation (1.13) maps the solution curve corresponding to a particular value of c_1 to the curve

$$\hat{y} = y = c_1 e^x = c_1 e^{\hat{x}-\varepsilon} = c_2 e^{\hat{x}}, \qquad \text{where} \quad c_2 = c_1 e^{-\varepsilon}.$$

Therefore translations in the x direction are nontrivial symmetries of (1.12), because (generally) $c_2 \neq c_1$. (Of course, $\varepsilon = 0$ necessarily gives a trivial symmetry.) Interestingly, one solution curve *is* mapped to itself by every translation, namely $y = 0$. Curves that are mapped to themselves by a symmetry are said to be *invariant* under the symmetry. The solution $y = 0$ partitions the set of solution curves $y = c_1 e^x$, as shown in Fig. 1.4. The translational symmetries

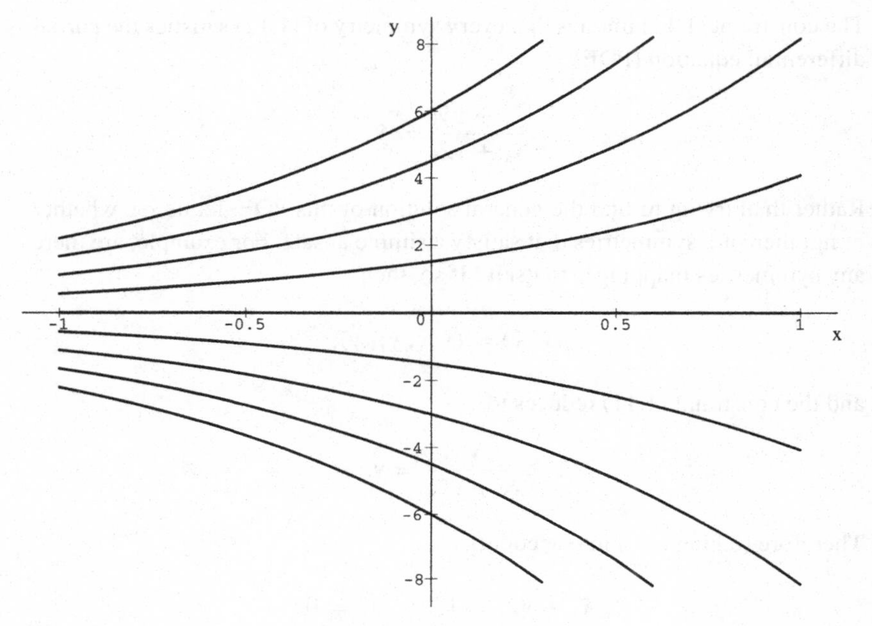

Fig. 1.4. Solutions of $y' = y$.

(1.13) are unable to map solutions with $c_1 > 0$ to solutions with $c_1 < 0$. However, the ODE does have symmetries that exchange the solutions in the upper and lower half-planes. One such symmetry is

$$(\hat{x}, \hat{y}) = (x, -y);$$

this is a discrete symmetry.

So far, we have looked at symmetries of very simple ODEs, but one strength of symmetry methods is that they are applicable to almost any ODE. Here are some more complicated examples.

Example 1.2 The Riccati equation

$$\frac{dy}{dx} = \frac{y+1}{x} + \frac{y^2}{x^3} \tag{1.14}$$

seems complicated, but its general solution is quite simple (as we shall see in the next chapter). The symmetries of this ODE include a one-parameter Lie group of inversions,

$$(\hat{x}, \hat{y}) = \left(\frac{x}{1-\varepsilon x}, \frac{y}{1-\varepsilon x}\right). \tag{1.15}$$

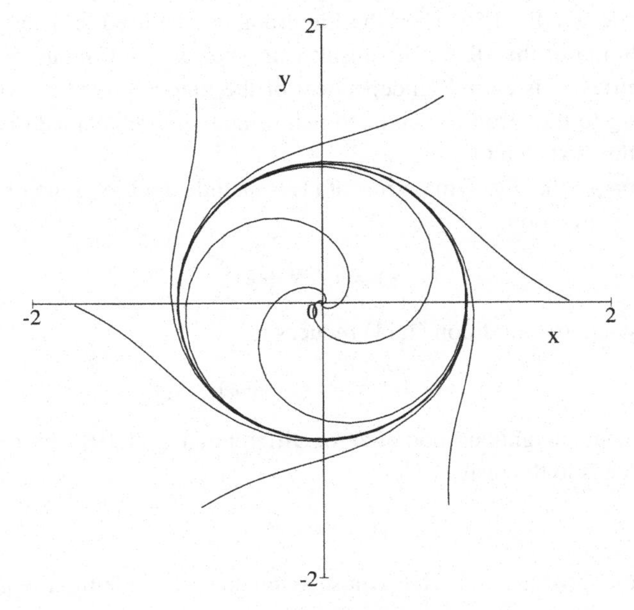

Fig. 1.5. Solutions of (1.16).

To prove this, simply substitute (1.15) into the symmetry condition (1.11), with ω defined by the right-hand side of (1.14). The inversions are our first example of a Lie group of symmetries that is not well defined for all real ε. (The radius of convergence of the Taylor series about $\varepsilon = 0$ is $1/|x|$.)

Example 1.3 Consider the ODE

$$\frac{dy}{dx} = \frac{y^3 + x^2 y - y - x}{xy^2 + x^3 + y - x}. \tag{1.16}$$

The set of solution curves is sketched in Fig. 1.5, which suggests that rotations about the origin are symmetries. It is left to the reader to check that the rotations

$$(\hat{x}, \hat{y}) = (x \cos \varepsilon - y \sin \varepsilon,\ x \sin \varepsilon + y \cos \varepsilon) \tag{1.17}$$

form a one-parameter Lie group of symmetries of (1.16).

1.4 Lie Symmetries Solve First-Order ODEs

The title of this section comes from the following rather surprising result. Suppose that we are able to find a nontrivial one-parameter Lie group of symmetries of a first-order ODE, (1.9). Then the Lie group can be used to determine the general solution of the ODE. This result is an indication of the usefulness of Lie symmetries; it is entirely independent of the function $\omega(x, y)$. The main ideas leading to the result are outlined below, and a more detailed discussion follows in the next chapter.

First, suppose that the symmetries of (1.9) include the Lie group of translations in the y direction,

$$(\hat{x}, \hat{y}) = (x, y + \varepsilon). \tag{1.18}$$

Then the symmetry condition (1.11) reduces to

$$\omega(x, y) = \omega(x, y + \varepsilon), \tag{1.19}$$

for all ε in some neighbourhood of zero. Differentiating (1.19) with respect to ε at $\varepsilon = 0$ leads to the result

$$\omega_y(x, y) = 0.$$

Therefore the most general ODE whose symmetries include the Lie group of translations (1.18) is of the form

$$\frac{dy}{dx} = \omega(x).$$

This ODE can be solved immediately: the general solution is

$$y = \int \omega(x)\, dx + c. \tag{1.20}$$

(We shall regard a differential equation as being solved if all that remains is to carry out *quadrature*, i.e., to evaluate an integral.) The particular solution corresponding to $c = 0$ is mapped by the translation to the solution

$$\hat{y} = \int \omega(x)\, dx + \varepsilon = \int \omega(\hat{x})\, d\hat{x} + \varepsilon,$$

which is the solution corresponding to $c = \varepsilon$. So by using the one-parameter Lie group, we obtain the general solution from one particular solution. The Lie group acts on the set of solution curves by changing the constant of integration.

Clearly, every first-order ODE with the Lie group of translations (1.18) is easily solved. Is the same true for ODEs with other one-parameter Lie groups? Consider the rotationally symmetric ODE (1.16), depicted in Fig. 1.5. It is natural to rewrite the ODE in terms of polar coordinates (r, θ), where

$$x = r\cos\theta, \qquad y = r\sin\theta.$$

We obtain a far simpler ODE,

$$\frac{dr}{d\theta} = r(1 - r^2), \tag{1.21}$$

which is immediately integrable. The one-parameter Lie group of rotations (1.17), rewritten in polar coordinates, becomes

$$(\hat{r}, \hat{\theta}) = (r, \theta + \varepsilon).$$

In the new coordinates, the rotational symmetries become translations in θ, so the ODE is easily solved.

The same idea works for all one-parameter Lie groups. In a suitable coordinate system, the symmetries parametrized by ε sufficiently close to zero are equivalent to translations (except at fixed points). One problem remains: what is the "suitable" coordinate system? For instance, the appropriate coordinate system for the ODE (1.14) is not obvious. It turns out that the Lie group itself holds the solution to this problem, as we shall see in the next chapter.

Exercises

1.1 Sketch the set of solutions of the ODE

$$\frac{dy}{dx} = \frac{y}{x}.$$

How many different kinds of symmetries can you identify?

1.2 Show that the transformation defined by

$$(\hat{x}, \hat{y}) = (e^{\varepsilon} x, y)$$

is a symmetry of

$$\frac{dy}{dx} = \frac{1 - y^2}{x}$$

for all $\varepsilon \in \mathbb{R}$. Describe these symmetries geometrically; how do they transform the solutions of the ODE?

1.3 Verify that the rotations (1.17) are symmetries of the ODE (1.16).

1.4 Determine the value of α for which

$$(\hat{x}, \hat{y}) = (x + 2\varepsilon, y e^{\alpha\varepsilon})$$

is a symmetry of

$$\frac{dy}{dx} = y^2 e^{-x} + y + e^x$$

for all $\varepsilon \in \mathbb{R}$.

1.5 Show that, for every $\varepsilon \in \mathbb{R}$,

$$(\hat{x}, \hat{y}) = \left(x, y + \varepsilon \exp\left\{\int F(x)\,dx\right\}\right)$$

is a symmetry of the general first-order linear ODE

$$\frac{dy}{dx} = F(x)y + G(x).$$

Explain the connection between these symmetries and the linear superposition principle.

2

Lie Symmetries of First-Order ODEs

Great floods have flown
From simple sources

(William Shakespeare: All's Well that Ends Well)

2.1 The Action of Lie Symmetries on the Plane

So far, we have considered only a few particular first-order ODEs of the form

$$\frac{dy}{dx} = \omega(x, y). \tag{2.1}$$

The purpose of this chapter is to develop techniques that are applicable to any ODE (2.1). We begin with a close examination of the way in which symmetries act on the plane. The main ideas are not difficult and can be illustrated with the aid of some very simple ODEs. Nevertheless, these ideas are quite general, and by the end of the chapter we shall have used them to solve ODEs that cannot be solved by standard methods.

Suppose that $y = f(x)$ is a solution of (2.1) and that a particular symmetry maps this solution to the curve $\hat{y} = \tilde{f}(\hat{x})$, which is a solution of

$$\frac{d\hat{y}}{d\hat{x}} = \omega(\hat{x}, \hat{y}).$$

The function $\tilde{f}$ is obtained as follows. The symmetry transforms the curve $y = f(x)$ to the set of points $(\hat{x}, \hat{y})$, where

$$\hat{x} = \hat{x}(x, f(x)), \qquad \hat{y} = \hat{y}(x, f(x)). \tag{2.2}$$

This is a curve in the $(\hat{x}, \hat{y})$ plane, written in parametric form (x is the parameter). Now solve the first equation of (2.2) to obtain x as a function of $\hat{x}$, and substitute the result into the second equation of (2.2). This gives

$$\tilde{f}(\hat{x}) = \hat{y}(x(\hat{x}), f(x(\hat{x}))). \tag{2.3}$$

If the symmetry belongs to a one-parameter Lie group, then $\tilde{f}$ is a function of $\hat{x}$ and the parameter ε.

Example 2.1 The general solution of the ODE

$$\frac{dy}{dx} = \frac{2y}{x} \tag{2.4}$$

is

$$y = cx^2. \tag{2.5}$$

Let us restrict attention to the quadrant $x > 0$, $y > 0$, in which each solution curve (2.5) corresponds to a particular $c > 0$. The set of solutions in this region is mapped to itself by the discrete symmetry

$$(\hat{x}, \hat{y}) = \left(\frac{x}{y}, \frac{1}{y}\right). \tag{2.6}$$

Specifically, the solution curve corresponding to $c = c_1$ is mapped to the curve

$$(\hat{x}, \hat{y}) = \left(\frac{1}{c_1 x}, \frac{1}{c_1 x^2}\right).$$

Therefore $x = 1/(c_1 \hat{x})$ and so the solution curve $y = c_1 x^2$ is mapped to

$$\hat{y} = c_1 \hat{x}^2.$$

The ODE (2.4) has many other symmetries, including the one-parameter Lie group of scalings

$$(\hat{x}, \hat{y}) = (e^{\varepsilon} x, e^{-\varepsilon} y). \tag{2.7}$$

Any symmetry of this form maps the solution curve $y = c_1 x^2$ to the curve

$$(\hat{x}, \hat{y}) = (e^{\varepsilon} x, c_1 e^{-\varepsilon} x^2).$$

Solving for x, we obtain $x = e^{-\varepsilon} \hat{x}$, and therefore the transformed solution is

$$\hat{y} = c_1 e^{-3\varepsilon} \hat{x}^2.$$

The $(\hat{x}, \hat{y})$ plane and the (x, y) plane contain the same set of solution curves. Instead of working with two identical planes, it is more convenient to superimpose them. Then the symmetry is regarded as a mapping of the (x, y) plane to itself, called the *action* of the symmetry on the (x, y) plane. Specifically, the point with coordinates (x, y) is mapped to the point whose coordinates are

$$(\hat{x}, \hat{y}) = (\hat{x}(x, y),\ \hat{y}(x, y)).$$

The solution curve $y = f(x)$ is the set of points with coordinates $(x, f(x))$. It is mapped to the set of points with coordinates $(\hat{x}, \tilde{f}(\hat{x}))$, that is, to the solution curve $y = \tilde{f}(x)$. Therefore the curve $y = f(x)$ is *invariant* under the symmetry if $\tilde{f} = f$. A symmetry is trivial if its action leaves every solution curve invariant.
(N.B. It may be necessary to restrict attention to a subset of the plane, if the ODE or the symmetry is not well defined on the whole plane.)

Example 2.2 In the previous example, we considered various symmetries of the ODE

$$\frac{dy}{dx} = \frac{2y}{x}, \qquad x > 0,\ y > 0.$$

We found that symmetries of the form (2.7) map the solution curve $y = c_1 x^2$ to the curve $\hat{y} = c_1 e^{-3\varepsilon} \hat{x}^2$ in the $(\hat{x}, \hat{y})$ plane. Therefore the action of a symmetry (2.7) on the quadrant $x > 0$, $y > 0$ maps the solution $y = c_1 x^2$ to the solution $y = c_1 e^{-3\varepsilon} x^2$, as shown in Fig. 2.1. The discrete symmetry (2.6) maps every solution curve $y = c_1 x^2$ to itself; hence (2.6) is a trivial symmetry.

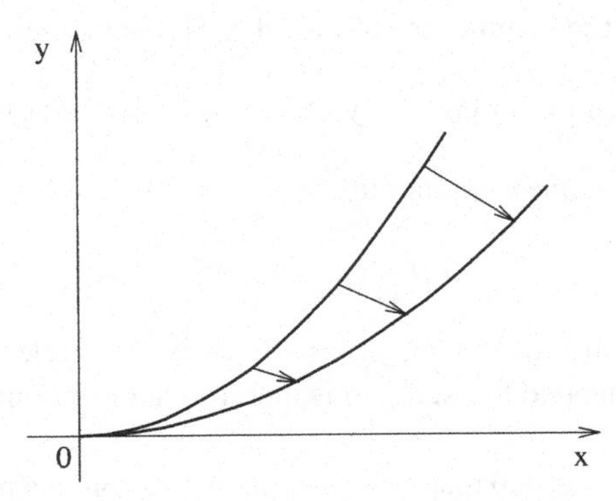

Fig. 2.1. The action of a symmetry (2.7) on a solution of (2.4).

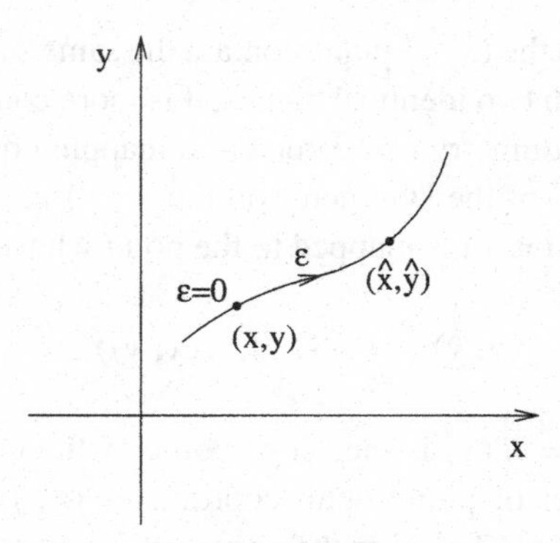

Fig. 2.2. Part of a one-dimensional orbit.

It is useful to study the action of a one-parameter Lie group of symmetries on points in the plane. The *orbit* of the group through (x, y) is the set of points to which (x, y) can be mapped by a suitable choice of ε. The coordinates of the points on the orbit through (x, y) are

$$(\hat{x}, \hat{y}) = (\hat{x}(x, y; \varepsilon),\ \hat{y}(x, y; \varepsilon)), \tag{2.8}$$

where

$$(\hat{x}(x, y; 0),\ \hat{y}(x, y; 0)) = (x, y). \tag{2.9}$$

The orbit through a typical point is a smooth curve, as shown in Fig. 2.2. However there may also be one or more *invariant points*, each of which is mapped to itself by the Lie symmetries. An invariant point is a zero-dimensional orbit of the Lie group.

Example 2.3 The Lie symmetries of the ODE (1.16) include the rotations

$$(\hat{x}, \hat{y}) = (x \cos\varepsilon - y \sin\varepsilon,\ x \sin\varepsilon + y \cos\varepsilon).$$

In polar coordinates, these amount to

$$(\hat{r}, \hat{\theta}) = (r,\ \theta + \varepsilon).$$

The orbit through any point $(x_0, y_0) \neq (0, 0)$ is the circle $r = \sqrt{x_0^2 + y_0^2}$, whereas $(0, 0)$ is mapped to itself and is therefore an invariant point.

The action of a Lie group maps every point on an orbit to a point on the same orbit. In other words, every orbit is invariant under the action of the Lie group.

Now consider the orbit through a noninvariant point (x, y). The *tangent vector* to the orbit at the point $(\hat{x}, \hat{y})$ is $(\xi(\hat{x}, \hat{y}), \eta(\hat{x}, \hat{y}))$, where

$$\frac{d\hat{x}}{d\varepsilon} = \xi(\hat{x}, \hat{y}), \qquad \frac{d\hat{y}}{d\varepsilon} = \eta(\hat{x}, \hat{y}). \tag{2.10}$$

In particular, the tangent vector at (x, y) is

$$\big(\xi(x, y), \eta(x, y)\big) = \left(\left.\frac{d\hat{x}}{d\varepsilon}\right|_{\varepsilon=0}, \left.\frac{d\hat{y}}{d\varepsilon}\right|_{\varepsilon=0} \right). \tag{2.11}$$

Therefore, to first order in ε, the Taylor series for the Lie group action is

$$\begin{aligned} \hat{x} &= x + \varepsilon\xi(x, y) + O(\varepsilon^2), \\ \hat{y} &= y + \varepsilon\eta(x, y) + O(\varepsilon^2). \end{aligned} \tag{2.12}$$

An invariant point is mapped to itself by every Lie symmetry. Therefore, from (2.12), the point (x, y) is invariant only if the tangent vector is zero, that is,

$$\xi(x, y) = \eta(x, y) = 0. \tag{2.13}$$

In fact this necessary condition is also sufficient, which can be proved by repeatedly differentiating (2.10) with respect to ε, then setting ε to zero. The set of tangent vectors for a particular Lie group is an example of a smooth *vector field*, because the tangent vectors vary smoothly with (x, y).

It is helpful to think of (2.10) as describing a steady flow of particles on the plane. In this analogy, ε is the "time" and the tangent vector at a point is the velocity of a particle at that point; the orbit is the pathline of the particle. Invariant points are the fixed points of the flow.

If an orbit crosses any curve C transversely at a point (x, y) then there are Lie symmetries that map (x, y) to points that are not on C. Therefore a curve is invariant if and only if no orbit crosses it. (The "if" holds because each orbit is invariant.) In other words, C is an invariant curve if and only if the tangent to C at each point (x, y) is parallel to the tangent vector $(\xi(x, y), \eta(x, y))$. This condition can be expressed mathematically by introducing the *characteristic*,

$$Q(x, y, y') = \eta(x, y) - y'\xi(x, y). \tag{2.14}$$

If C is the curve $y = y(x)$, the tangent to C at $(x, y(x))$ is in the direction $(1, y'(x))$; it is parallel to $(\xi(x, y), \eta(x, y))$ if and only if

$$Q(x, y, y') = 0 \qquad \text{on } C. \tag{2.15}$$

This result enables us to characterize the invariant solutions of (2.1), as follows. On all solutions of (2.1), the characteristic is equivalent to

$$\bar{Q}(x, y) = Q(x, y, \omega(x, y)) = \eta(x, y) - \omega(x, y)\,\xi(x, y). \tag{2.16}$$

[We call $\bar{Q}(x, y)$ the *reduced characteristic*.] A solution curve $y = f(x)$ is invariant if and only if

$$\bar{Q}(x, y) = 0 \qquad \text{when} \quad y = f(x). \tag{2.17}$$

The Lie symmetries are trivial if and only if $\bar{Q}(x, y)$ is identically zero, that is,

$$\eta(x, y) \equiv \omega(x, y)\xi(x, y). \tag{2.18}$$

If $\bar{Q}_y \not\equiv 0$ then it is possible to determine the curves $y = f(x)$ that satisfy (2.17). Every such curve is an invariant solution of (2.1), as will be shown later. Therefore (2.17) can be used to find all solutions that are invariant under a given nontrivial Lie group, without it being necessary to carry out any integration!

Example 2.4 The ODE

$$\frac{dy}{dx} = y \tag{2.19}$$

has scaling symmetries of the form

$$(\hat{x}, \hat{y}) = (x,\ e^{\varepsilon} y). \tag{2.20}$$

The tangent vector at (x, y) is found by differentiating (2.20) with respect to ε at $\varepsilon = 0$:

$$(\xi(x, y),\ \eta(x, y)) = (0, y). \tag{2.21}$$

From (2.13), every point on the line $y = 0$ is invariant under the action of the Lie symmetries (2.20). On the solutions of (2.19), the characteristic reduces to

$$\bar{Q}(x, y) = \eta(x, y) - y\xi(x, y) = y.$$

Therefore this Lie group acts nontrivially on the solutions of (2.19). From (2.17), the only invariant solution is $y = 0$, which is composed entirely of invariant points. Here is another one-parameter Lie group of symmetries of (2.19):

$$(\hat{x}, \hat{y}) = (e^{\varepsilon} x,\ \exp\{(e^{\varepsilon} - 1)x\}y). \tag{2.22}$$

The tangent vector at (x, y) is

$$(\xi(x, y),\ \eta(x, y)) = (x, xy). \tag{2.23}$$

Every point on the line $x = 0$ is invariant. Furthermore,

$$\bar{Q}(x, y) = \eta(x, y) - y\xi(x, y) = 0,$$

so the Lie symmetries (2.22) act trivially on the solutions of (2.19).

Example 2.5 The Riccati equation

$$y' = xy^2 - \frac{2y}{x} - \frac{1}{x^3}, \qquad (x \neq 0) \tag{2.24}$$

has a Lie group of scaling symmetries

$$(\hat{x}, \hat{y}) = (e^{\varepsilon}x, e^{-2\varepsilon}y).$$

The tangent vector field is

$$(\xi(x, y),\ \eta(x, y)) = (x,\ -2y),$$

and so the reduced characteristic is

$$\bar{Q}(x, y) = \frac{1}{x^2} - x^2y^2.$$

Therefore the Lie symmetries are nontrivial, and there are two invariant solutions:

$$y = \pm x^{-2}.$$

Most symmetry methods use the tangent vectors, rather than the symmetries themselves. However, Lie symmetries can be reconstructed from the tangent vectors by integrating the coupled ODEs (2.10) subject to the initial conditions (2.9). So (locally) there is a one-to-one correspondence between each one-parameter Lie group and its tangent vector field.

Example 2.6 The Lie symmetries associated with the tangent vector field (2.21) are reconstructed as follows. Substitute (2.21) into (2.10) to obtain

$$\frac{d\hat{x}}{d\varepsilon} = 0, \qquad \frac{d\hat{y}}{d\varepsilon} = \hat{y},$$

whose general solution is

$$\hat{x}(x, y; \varepsilon) = A(x, y), \qquad \hat{y}(x, y; \varepsilon) = B(x, y)e^{\varepsilon}.$$

Then set $\varepsilon = 0$ and use the initial condition (2.9), which yields

$$(\hat{x}, \hat{y}) = (x, e^{\varepsilon} y)$$

in agreement with (2.20).

2.2 Canonical Coordinates

In §1.4 we found that every ODE (2.1) whose symmetries include the translations

$$(\hat{x}, \hat{y}) = (x, y + \varepsilon) \tag{2.25}$$

may be integrated directly. More generally, if an ODE has Lie symmetries that are equivalent to translations (under a change of coordinates), the ODE can be solved by rewriting it in terms of the new coordinates. How can these coordinates be found?

All orbits of the symmetries (2.25) have the same tangent vector at every point:

$$(\xi(x, y), \eta(x, y)) = (0, 1). \tag{2.26}$$

[The orbits of (2.25) are the lines of constant x.] Given any one-parameter Lie group of symmetries, we aim to introduce coordinates

$$(r, s) = \big(r(x, y), s(x, y)\big)$$

such that

$$(\hat{r}, \hat{s}) \equiv \big(r(\hat{x}, \hat{y}), s(\hat{x}, \hat{y})\big) = (r, s + \varepsilon). \tag{2.27}$$

If this is possible then, in the new coordinates, the tangent vector at the point (r, s) is $(0, 1)$, that is,

$$\left.\frac{d\hat{r}}{d\varepsilon}\right|_{\varepsilon=0} = 0, \qquad \left.\frac{d\hat{s}}{d\varepsilon}\right|_{\varepsilon=0} = 1.$$

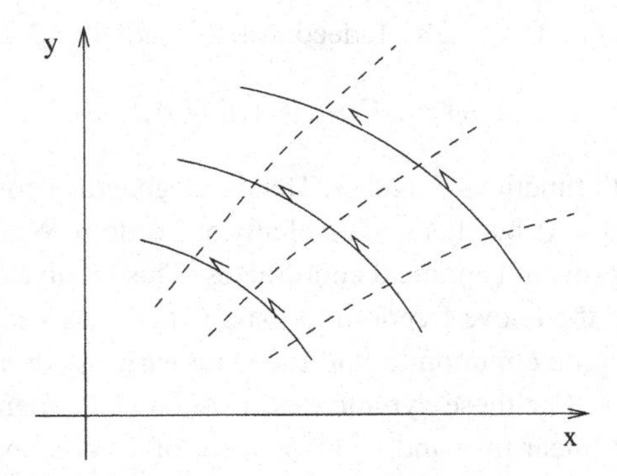

Fig. 2.3. Curves of constant r (—) and s (- - -). Some tangent vectors are shown.

Using the chain rule and (2.10), we obtain

$$\begin{aligned}\xi(x, y)r_x + \eta(x, y)r_y &= 0,\\ \xi(x, y)s_x + \eta(x, y)s_y &= 1.\end{aligned} \tag{2.28}$$

The change of coordinates should be invertible in some neighbourhood of (x, y), so we impose the nondegeneracy condition

$$r_x s_y - r_y s_x \neq 0. \tag{2.29}$$

This condition ensures that if a curve of constant s and a curve of constant r meet at a point, they cross one another transversely, as shown in Fig. 2.3. Any pair of functions $r(x, y)$, $s(x, y)$ satisfying (2.28) and (2.29) is called a pair of *canonical coordinates*.

By definition, the tangent vector at any noninvariant point is parallel to the curve of constant r passing through that point. Therefore the curve of constant r coincides (locally) with the orbit through the point. The orbit is invariant under the Lie group, so r is sometimes referred to as an *invariant canonical coordinate*. The curves of constant s are not invariant, because they cross the one-dimensional orbits transversely.

Canonical coordinates cannot be defined at an invariant point, because the determining equation for s in (2.28) has no solution if

$$\xi(x, y) = \eta(x, y) = 0.$$

However, canonical coordinates do exist in some neighbourhood of any non-invariant point. In other words, it is always possible to normalise the tangent vectors (at least, locally) provided that they are nonzero. Canonical coordinates are not uniquely defined by (2.28). Indeed, if (r, s) satisfies (2.28), then so does

$$(\tilde{r}, \tilde{s}) = (F(r),\ s + G(r)), \tag{2.30}$$

for arbitrary smooth functions F and G. The nondegeneracy condition imposes the constraint $F'(r) \neq 0$, but there is still plenty of freedom. We intend to rewrite the ODE (2.1) in terms of canonical coordinates. This involves differentiation, so it is wise to use the above freedom to make r and s as simple as possible. For example, it is quite common to find Lie symmetries with η linear in y and ξ independent of y. For these symmetries, if $\xi(x) \neq 0$, there are canonical coordinates with r linear in y and s independent of y. Wherever possible, we shall try to use a simple nondegenerate solution of (2.28).

Canonical coordinates can be obtained from (2.28) by using the method of characteristics. The characteristic equations are

$$\frac{dx}{\xi(x, y)} = \frac{dy}{\eta(x, y)} = ds. \tag{2.31}$$

A *first integral* of a given first-order ODE

$$\frac{dy}{dx} = f(x, y) \tag{2.32}$$

is a nonconstant function $\phi(x, y)$ whose value is constant on any solution $y = y(x)$ of the ODE (2.32). Therefore

$$\phi_x + f(x, y)\phi_y = 0, \qquad \phi_y \neq 0. \tag{2.33}$$

The general solution of (2.32) is

$$\phi(x, y) = c. \tag{2.34}$$

Suppose that $\xi(x, y) \neq 0$. Comparing (2.28) and (2.33), we see that the invariant canonical coordinate r is a first integral of

$$\frac{dy}{dx} = \frac{\eta(x, y)}{\xi(x, y)}. \tag{2.35}$$

So $r = \phi(x, y)$ is found by solving (2.35). Quite often, a solution $s(x, y)$ of (2.28) may be found by inspection. Otherwise we can use $r = r(x, y)$ to write y as a function of r and x. Then the coordinate $s(r, x)$ is obtained from (2.31) by quadrature:

$$s(r, x) = \left(\int \frac{dx}{\xi(x, y(r, x))} \right) \bigg|_{r=r(x,y)}; \tag{2.36}$$

here the integral is evaluated with r being treated as a constant.

Similarly, if $\xi(x, y) = 0$ and $\eta(x, y) \neq 0$ then

$$r = x, \qquad s = \left(\int \frac{dy}{\eta(r, y)} \right) \Bigg|_{r=x}, \tag{2.37}$$

are canonical coordinates.

Example 2.7 Consider the following Lie symmetries, which are scalings:

$$(\hat{x}, \hat{y}) = (e^{\varepsilon} x, e^{k\varepsilon} y), \qquad k > 0. \tag{2.38}$$

The tangent vector is

$$(\xi(x, y), \eta(x, y)) = (x, ky),$$

and therefore r is a first integral of

$$\frac{dy}{dx} = \frac{ky}{x}.$$

The general solution of this ODE is $y = cx^k$, so we choose $r = x^{-k} y$, for simplicity. As ξ is nonzero and independent of y, it is easiest to choose s to be a function of x alone. We obtain

$$(r, s) = (x^{-k} y, \ln |x|). \tag{2.39}$$

These canonical coordinates cannot be used on the whole plane: $s = \ln |x|$ fails on the line $x = 0$. The following canonical coordinates are suitable for use near to $x = 0$, except on the line $y = 0$:

$$(r, s) = (x^k y^{-1}, k^{-1} \ln |y|). \tag{2.40}$$

Canonical coordinates do not exist at the invariant point $(0, 0)$.

The above example illustrates a minor difficulty with canonical coordinates. They cannot be defined at an invariant point, and so it is necessary to use several coordinate "patches" to cover all noninvariant points.

Example 2.8 The one-parameter Lie group of inversions

$$(\hat{x}, \hat{y}) = \left(\frac{x}{1 - \varepsilon x}, \frac{y}{1 - \varepsilon x} \right) \tag{2.41}$$

has the tangent vector field

$$(\xi(x, y), \eta(x, y)) = (x^2, xy). \tag{2.42}$$

The method outlined above yields the following (simple) canonical coordinates:

$$(r, s) = \left(\frac{y}{x}, -\frac{1}{x}\right) \qquad x \neq 0. \tag{2.43}$$

Every point on the line $x = 0$ is invariant, so canonical coordinates cannot be defined there.

The most common use of canonical coordinates is for obtaining solutions of ODEs. However, many features of Lie symmetries become transparent when canonical coordinates are used. For example, it becomes easy to reconstruct the Lie symmetries, as follows. First, write x and y in terms of the canonical coordinates:

$$x = f(r, s), \qquad y = g(r, s).$$

Therefore, from (2.27),

$$\begin{aligned} \hat{x} &= f(\hat{r}, \hat{s}) = f(r(x, y), s(x, y) + \varepsilon), \\ \hat{y} &= g(\hat{r}, \hat{s}) = g(r(x, y), s(x, y) + \varepsilon). \end{aligned} \tag{2.44}$$

Example 2.9 Consider the tangent vector (2.42); the canonical coordinates (2.43) are inverted to obtain

$$(x, y) = \left(-\frac{1}{s}, -\frac{r}{s}\right).$$

Therefore (2.44) gives

$$(\hat{x}, \hat{y}) = \left(-\frac{1}{s+\varepsilon}, -\frac{r}{s+\varepsilon}\right) = \left(\frac{x}{1-\varepsilon x}, \frac{y}{1-\varepsilon x}\right),$$

as expected.

2.3 How to Solve ODEs with Lie Symmetries

Suppose that we have been able to find nontrivial Lie symmetries of a given ODE (2.1). Recall that Lie symmetries are nontrivial if and only if

$$\eta(x, y) \not\equiv \omega(x, y)\xi(x, y). \tag{2.45}$$

(The reason for this restriction is discussed later.)

Then the ODE (2.1) can be reduced to quadrature by rewriting it in terms of canonical coordinates as follows:

$$\frac{ds}{dr} = \frac{s_x + \omega(x, y)s_y}{r_x + \omega(x, y)r_y}. \tag{2.46}$$

The right-hand side of (2.46) can now be written as a function of r and s. For a general change of variables $(x, y) \mapsto (r, s)$, the transformed ODE would be of the form

$$\frac{ds}{dr} = \Omega(r, s), \tag{2.47}$$

for some function Ω. However, (r, s) are canonical coordinates, and so the ODE is invariant under the group of translations in the s direction:

$$(\hat{r}, \hat{s}) = (r,\ s + \varepsilon).$$

Therefore, from §1.4, the ODE (2.47) is of the form

$$\frac{ds}{dr} = \Omega(r). \tag{2.48}$$

The problem is now reduced to quadrature. The general solution of (2.48) is

$$s - \int \Omega(r)\, dr = c,$$

where c is an arbitrary constant. Therefore the general solution of the original ODE (2.1) is

$$s(x, y) - \int^{r(x,y)} \Omega(r)\, dr = c. \tag{2.49}$$

This very simple method can be applied to any ODE (2.1) with a known nontrivial one-parameter Lie group of symmetries. Of course, one must first determine the canonical coordinates by solving the ODE (2.35). Typically, (2.35) is very much easier to solve than (2.1). The examples below demonstrate the effectiveness of the method in dealing with ODEs whose solutions are not obvious. (Henceforth, arbitrary constants will be denoted by c or c_i.)

Example 2.10 We have already found the solutions of the Riccati equation

$$y' = xy^2 - \frac{2y}{x} - \frac{1}{x^3}, \qquad (x \neq 0), \tag{2.50}$$

that are invariant under the Lie symmetries

$$(\hat{x}, \hat{y}) = (e^{\varepsilon} x,\ e^{-2\varepsilon} y).$$

Let us now complete the solution of (2.50). From (2.39), suitable canonical coordinates are

$$(r, s) = (x^2 y,\ \ln|x|).$$

Then (2.50) reduces to

$$\frac{ds}{dr} = \frac{1}{r^2 - 1}.$$

The quadrature is straightforward, and we find (after writing r and s in terms of x and y) that the general solution of (2.50) is

$$y = \frac{c + x^2}{x^2(c - x^2)}. \tag{2.51}$$

The invariant solution curve $y = x^{-2}$ can be regarded as the limit of (2.51) as c approaches infinity. The other invariant solution is obtained by setting $c = 0$ in (2.51).

Example 2.11 In Example 1.2, we found that the ODE

$$\frac{dy}{dx} = \frac{y+1}{x} + \frac{y^2}{x^3}$$

has Lie symmetries of the form

$$(\hat{x}, \hat{y}) = \left(\frac{x}{1 - \varepsilon x},\ \frac{y}{1 - \varepsilon x} \right).$$

The canonical coordinates for these inversions are given by (2.43). They reduce the ODE to

$$\frac{ds}{dr} = \frac{1}{1 + r^2}.$$

The general solution is

$$s = \tan^{-1}(r) + c,$$

which is equivalent to

$$y = -x \tan\left(\frac{1}{x} + c \right).$$

Example 2.12 The ODE

$$y' = \frac{y - 4xy^2 - 16x^3}{y^3 + 4x^2y + x} \tag{2.52}$$

has Lie symmetries whose tangent vector field is

$$(\xi(x, y),\ \eta(x, y)) = (-y, 4x).$$

(A method for deriving this vector field directly from the ODE is outlined in §2.4.) The characteristic equation for r is

$$\frac{dy}{dx} = -\frac{4x}{y},$$

and so we can take $r = \sqrt{4x^2 + y^2}$. Now consider the region $y > 0$; here $y(r, x) = \sqrt{r^2 - 4x^2}$, and so a second canonical coordinate is

$$s = -\int^x \frac{dx}{\sqrt{r^2 - 4x^2}} = \frac{1}{2}\cos^{-1}\left(\frac{2x}{r}\right) = \frac{1}{2}\cot^{-1}\left(\frac{2x}{y}\right), \qquad s \in (0, \pi/2).$$

In this region, the ODE (2.52) reduces to

$$\frac{ds}{dr} = -r.$$

[The reduction to quadrature is similar in other regions of the (x, y) plane.] Reverting to the original variables, we obtain the general solution of (2.52):

$$y\cos(4x^2 + y^2 + c) + 2x\sin(4x^2 + y^2 + c) = 0.$$

Why is it necessary to exclude one-parameter Lie groups whose action on the set of solutions is trivial? In principle, there is no difficulty in defining canonical coordinates in the usual way. Suppose that (r, s) are canonical coordinates for a trivial one-parameter Lie group of symmetries of a given ODE (2.1). Write the general solution of the ODE as

$$\phi(r, s) = c.$$

Every solution is invariant under the action of the Lie symmetries, and therefore

$$\phi(r, s + \varepsilon) = \phi(r, s),$$

for all ε sufficiently close to zero. Hence ϕ is independent of s and, without loss of generality, the general solution of the ODE can be rewritten as

$$r = c.$$

So we need only find the invariant canonical coordinate r, which is a first integral of

$$\frac{dy}{dx} = \frac{\eta(x, y)}{\xi(x, y)}. \tag{2.53}$$

Lie symmetries of (2.1) are trivial if and only if

$$\eta(x, y) \equiv \omega(x, y)\xi(x, y),$$

and so (2.53) reduces to the original ODE (2.1)! Our aim is to solve this ODE, so the trivial symmetries are useless.

2.4 The Linearized Symmetry Condition

How can one find symmetries of (2.1)? One method is to use the symmetry condition (1.10), which is equivalent to

$$\frac{\hat{y}_x + \omega(x, y)\hat{y}_y}{\hat{x}_x + \omega(x, y)\hat{x}_y} = \omega(\hat{x}, \hat{y}). \tag{2.54}$$

In general, this is a complicated nonlinear partial differential equation in the two unknowns $\hat{x}$ and $\hat{y}$. However, Lie symmetries can be derived from a much simpler condition on the tangent vector field. (Remember that once the tangent vectors have been found, the Lie symmetries can be reconstructed.)

By definition, the Lie symmetries of (2.1) are of the form

$$\begin{aligned} \hat{x} &= x + \varepsilon\xi(x, y) + O(\varepsilon^2), \\ \hat{y} &= y + \varepsilon\eta(x, y) + O(\varepsilon^2), \end{aligned} \tag{2.55}$$

for some smooth functions ξ and η. To simplify notation, the arguments (x, y) will be omitted from ξ and η from now on. Substituting (2.55) into (2.54), we obtain

$$\frac{\omega(x, y) + \varepsilon\{\eta_x + \omega(x, y)\eta_y\} + O(\varepsilon^2)}{1 + \varepsilon\{\xi_x + \omega(x, y)\xi_y\} + O(\varepsilon^2)} = \omega(x + \varepsilon\xi + O(\varepsilon^2), y + \varepsilon\eta + O(\varepsilon^2)). \tag{2.56}$$

We now expand each side of (2.56) as a Taylor series about $\varepsilon = 0$, assuming that each series converges.

$$\omega + \varepsilon\{\eta_x + (\eta_y - \xi_x)\omega - \xi_y\omega^2\} + O(\varepsilon^2) = \omega + \varepsilon\{\xi\omega_x + \eta\omega_y\} + O(\varepsilon^2);$$

here ω is shorthand for $\omega(x, y)$. This condition is necessarily satisfied at $\varepsilon = 0$, which corresponds to the trivial symmetery $(\hat{x}, \hat{y}) = (x, y)$. Equating the $O(\varepsilon)$ terms gives the *linearized symmetry condition*

$$\eta_x + (\eta_y - \xi_x)\omega - \xi_y\omega^2 = \xi\omega_x + \eta\omega_y. \tag{2.57}$$

Like (2.54), this is a single PDE involving two dependent variables, which has infinitely many functionally independent solutions. However, (2.57) is linear and is simpler than (2.54). It is usually much easier to find solutions of (2.57) using some ansatz than to try to solve (2.54) directly.

The linearized symmetry condition can be rewritten in terms of the reduced characteristic

$$\bar{Q} = \eta - \omega\xi,$$

as follows:

$$\bar{Q}_x + \omega\bar{Q}_y = \omega_y\bar{Q}. \tag{2.58}$$

Each solution of (2.58) corresponds to infinitely many Lie groups, for if $\bar{Q}$ satisfies (2.58) then

$$(\xi, \eta) = (\xi,\ \bar{Q} + \omega\xi)$$

is a tangent vector field of a one-parameter group, for any function ξ. All trivial Lie symmetries correspond to the solution $\bar{Q} = 0$ of (2.58). In principle, the nontrivial symmetries can be found from (2.58) by using the method of characteristics. The characteristic equations are

$$\frac{dx}{1} = \frac{dy}{\omega(x, y)} = \frac{d\bar{Q}}{\omega_y(x, y)\bar{Q}}. \tag{2.59}$$

The first equation of (2.59) is equivalent to the ODE (2.1), so usually one cannot find a nonzero solution of (2.58) without knowing the general solution of (2.1).

N.B. If (ξ, η) is a nonzero solution of (2.57), then so is $(k\xi, k\eta)$, for any nonzero constant, k. This freedom corresponds to replacing ε by $k^{-1}\varepsilon$, which does not alter the orbits of the Lie group. So the same Lie symmetries are recovered, irrespective of the value of k. The freedom to rescale ε means that

$\bar{Q}$ can be multiplied by any convenient nonzero constant, without affecting the orbits.

To solve (2.57) it is necessary to use an appropriate ansatz, that is, to place some additional constraints upon ξ and η. The following example illustrates the general idea.

Example 2.13 Consider the ODE

$$\frac{dy}{dx} = \frac{1-y^2}{xy} + 1. \tag{2.60}$$

Here the function $\omega(x, y)$ is fairly simple, so let us try an ansatz that is not too restrictive. Many Lie symmetries have tangent vector fields of the form

$$\xi = \alpha(x), \qquad \eta = \beta(x)y + \gamma(x),$$

for some functions α, β and γ. Does (2.60) have any such symmetries? If so, the linearized symmetry condition is

$$\beta' y + \gamma' + (\beta - \alpha')\left(\frac{1-y^2}{xy} + 1\right) = \alpha\left(\frac{y^2-1}{x^2 y}\right) - (\beta y + \gamma)\left(\frac{1+y^2}{xy^2}\right). \tag{2.61}$$

Although (2.61) is a single equation, it can be split into an overdetermined system of equations by comparing terms that are multiplied by each power of y. The y^{-2} terms give

$$\gamma = 0,$$

and the y^{-1} terms are

$$\frac{\beta - \alpha'}{x} = -\frac{\alpha}{x^2} - \frac{\beta}{x}.$$

The terms that are independent of y give

$$\beta = \alpha',$$

and hence

$$\alpha' + \frac{\alpha}{x} = 0.$$

This ODE is easily solved; the general solution is

$$\alpha = c_1 x^{-1},$$

and therefore

$$\beta = -c_1 x^{-2}.$$

The remaining terms in the linearized symmetry condition provide no further constraints, so any tangent vector field of the form

$$(\xi, \eta) = (c_1 x^{-1},\ -c_1 x^{-2} y)$$

satisfies the linearized symmetry condition.

Having found a tangent vector field, we can easily check whether it is trivial; if it is not, it may be used to solve the ODE. Generally speaking, the difficulty of the calculations increases with the generality of the ansatz. If $\omega(x, y)$ is complicated, it is a good idea to begin with a fairly restrictive ansatz in order to keep the calculations manageable. If this does not succeed, then one may need to resort to computer algebra in order to try out more general ansätze.

Example 2.14 Some common symmetries, including translations, scalings and rotations, can be found with the ansatz

$$\xi = c_1 x + c_2 y + c_3, \qquad \eta = c_4 x + c_5 y + c_6. \tag{2.62}$$

On the whole, (2.62) is more restrictive than the ansatz used in the previous example, for it has no arbitrary functions. The reader is invited to verify that (2.62) satisfies the linearized symmetry condition for the ODE (2.52) provided that

$$c_1 = c_3 = c_5 = c_6 = 0, \qquad c_4 = -4c_2.$$

(The use of a computer algebra package is recommended, though not essential, for this example.)

There are now several reliable computer algebra packages that can greatly reduce the effort involved in finding and using symmetries. For example, the *ODEtools* package within MapleV (Release 5) has a program called Symgen, which attempts to find solutions of (2.57) with an ansatz defined by the user.

Nevertheless, for some ODEs, the search for a nontrivial symmetry may be fruitless, even though infinitely many such symmetries exist. This is the chief obstacle to using symmetries of first-order ODEs. (Symmetries of higher-order ODEs and PDEs, however, *can* usually be found systematically.)

We end this section by justifying the assertion made earlier, that every curve $y = f(x)$ satisfying

$$\bar{Q} = 0 \quad \text{when} \quad y = f(x), \qquad \bar{Q}_y \neq 0, \tag{2.63}$$

is a solution of (2.1). Differentiate (2.63) with respect to x, to obtain

$$\bar{Q}_x + f'(x)\bar{Q}_y = 0 \quad \text{when} \quad y = f(x). \tag{2.64}$$

Now compare (2.58) and (2.64), taking (2.63) into account. This gives

$$f'(x) = \omega(x, f(x)),$$

which is the required result.

2.5 Symmetries and Standard Methods

Canonical coordinates are associated with a particular Lie group. So all first-order ODEs admitting a given one-parameter symmetry group can be reduced to quadrature, using canonical coordinates defined by the group generator. This gives rise to methods that work for whole classes of ODEs, some of which are taught as standard methods.

Example 2.15 Every ODE of the form

$$\frac{dy}{dx} = F\left(\frac{y}{x}\right) \tag{2.65}$$

admits the one-parameter Lie group of scaling symmetries

$$(\hat{x}, \hat{y}) = (e^{\varepsilon}x, e^{\varepsilon}y).$$

The standard solution method for this type of ODE is to introduce new variables

$$r = \frac{y}{x}, \qquad s = \ln|x|;$$

these are canonical coordinates (for $x \neq 0$). There are two possibilities. If $F(r) = r$, the symmetries are trivial, and the general solution of (2.65) is $r = c$, that is,

$$y = cx.$$

Otherwise, (2.65) is equivalent to

$$\frac{ds}{dr} = \frac{1}{F(r) - r},$$

so the general solution of (2.65) is

$$\ln|x| = \int^{y/x} \frac{dr}{F(r) - r} + c.$$

Example 2.16 Consider the general linear ODE

$$y' + F(x)y = G(x). \tag{2.66}$$

The homogeneous ODE

$$u' + F(x)u = 0$$

is separable; one nonzero solution is

$$u = u_0(x) \equiv \exp\left\{-\int F(x)\,dx\right\}.$$

The principle of *linear superposition* states that if $y = y(x)$ is a solution of (2.66), then so is $y = y(x) + \varepsilon u_0(x)$, for each $\varepsilon \in \mathbb{R}$. This principle is equivalent to the statement that (2.66) has the Lie symmetries

$$(\hat{x}, \hat{y}) = (x, y + \varepsilon u_0(x)).$$

The tangent vector field is

$$(\xi, \eta) = (0, u_0(x)),$$

so some simple canonical coordinates are

$$(r, s) = \left(x, \frac{y}{u_0(x)}\right).$$

In these coordinates, (2.66) is equivalent to

$$\frac{ds}{dr} = \frac{G(r)}{u_0(r)}.$$

Thus we obtain the well-known result that the general solution of (2.66) is

$$\frac{y}{u_0(x)} - \int^{x} \frac{G(r)}{u_0(r)}\,dr = c.$$

Similarly, almost all standard methods that use a change of variables are special cases of the general technique outlined above. A different approach is to use an *integrating factor* to solve the ODE (2.1), written in the "Pfaffian" form

$$dy - \omega\, dx = 0. \tag{2.67}$$

The aim is to find a function $\mu(x, y)$ such that the general solution of (2.67) can be written as a line integral:

$$\Phi(x, y) \equiv \int \mu(dy - \omega\, dx) = c.$$

Consequently

$$\Phi_x = -\omega\mu, \qquad \Phi_y = \mu,$$

which leads to the condition

$$\mu_x + (\omega\mu)_y = 0. \tag{2.68}$$

Comparing (2.68) with (2.58), we find that

$$\mu(x, y) = \frac{1}{\bar{Q}(x, y)}$$

is an integrating factor, provided that $\bar{Q}$ is not identically zero. Therefore if (ξ, η) is the tangent vector field for a nontrivial one-parameter Lie group of symmetries of (2.1), the general solution of (2.1) is

$$\int \frac{dy - \omega\, dx}{\eta - \omega\xi} = c. \tag{2.69}$$

The integrating factor method is entirely equivalent to the method of canonical coordinates. To see this, we rearrange (2.46) and (2.48) as follows:

$$\omega = -\frac{s_x - \Omega(r) r_x}{s_y - \Omega(r) r_y}. \tag{2.70}$$

From (2.28),

$$\xi\{s_x - \Omega(r) r_x\} + \eta\{s_y - \Omega(r) r_y\} = 1.$$

Therefore

$$s_x - \Omega(r) r_x = -\frac{\omega}{\eta - \omega\xi},$$
$$s_y - \Omega(r) r_y = \frac{1}{\eta - \omega\xi},$$

and so (2.69) is equivalent to

$$\int ds - \Omega(r)\,dr = \int \{s_x - \Omega(r) r_x\}\,dx + \{s_y - \Omega(r) r_y\}\,dy = c.$$

The integrating factor method is particularly useful if it is difficult to calculate canonical coordinates, as the following example shows.

Example 2.17 The ODE

$$y' = \frac{y^3 + y - 3x^2 y}{3xy^2 + x - x^3} \tag{2.71}$$

is not easily solved by any standard method, but is invariant under the symmetry group whose tangent vector field is

$$(\xi, \eta) = (y^3 + y - 3x^2 y,\ x^3 - x - 3xy^2).$$

The characteristic equation for $r(x, y)$ is

$$\frac{dy}{dx} = \frac{x^3 - x - 3xy^2}{y^3 + y - 3x^2 y},$$

which seems as difficult to solve as (2.71). Now let us try to use (2.69); this gives (after factorization)

$$\int \frac{(x^3 - x - 3xy^2)\,dy + (y^3 + y - 3x^2 y)\,dx}{(y^2 + x^2)(y^2 + (x+1)^2)(y^2 + (x-1)^2)} = c,$$

which is readily integrated (using partial fractions) to yield

$$\frac{1}{2}\tan^{-1}\left(\frac{y}{x+1}\right) + \frac{1}{2}\tan^{-1}\left(\frac{y}{x-1}\right) - \tan^{-1}\left(\frac{y}{x}\right) = c.$$

Standard trigonometric identities reduce this to the simpler form

$$\frac{(y^2 + x^2)^2 + y^2 - x^2}{xy} = c_1.$$

We have two different methods of solution to choose from; if one method presents problems, the other may succeed. The difficulty is that one must first find a nontrivial symmetry group, a task that may not be easy!

An alternative approach is to classify all ODEs that have a *given* Lie group. The idea is simple:

(i) choose a symmetry group, and use its generator to determine the canonical coordinates $r(x, y)$ and $s(x, y)$;

(ii) from (2.70), the most general first order ODE invariant under this group is

$$y' = -\frac{s_x(x, y) - \Omega(r(x, y))r_x(x, y)}{s_y(x, y) - \Omega(r(x, y))r_y(x, y)}, \tag{2.72}$$

where Ω is an arbitrary smooth function.

In this way, a catalogue of equation types can be constructed. The catalogue cannot be complete, for several reasons. There are infinitely many symmetry groups of first order ODEs, so a complete list would need an infinity of entries. Furthermore, the method relies on the construction of canonical coordinates; we have seen that this is not always easy. Nevertheless, it is helpful to be able to recognise some of the symmetries that occur most frequently in applications.

Example 2.18 The group of rotations about the origin has the tangent vector field

$$(\xi, \eta) = (-y, x).$$

The usual canonical coordinates for this Lie group are the polar coordinates

$$(r, s) = \left(\sqrt{x^2 + y^2},\ \tan^{-1}\left(\frac{y}{x}\right)\right).$$

So, from (2.72), the most general first order ODE invariant under the group of rotations is

$$y' = \frac{y + x\Omega(\sqrt{x^2 + y^2})}{x - y\Omega(\sqrt{x^2 + y^2})},$$

where Ω is an arbitrary smooth function.

2.6 The Infinitesimal Generator

So far, we have restricted attention to first-order ODEs of the form (2.1); this has enabled us to discuss many of the geometrical ideas that are the foundation of symmetry methods. We need to extend these ideas to higher-order ODEs and PDEs, and it will no longer be possible to use two-dimensional pictures to represent everything of importance. Instead, we introduce a compact notation that is easily extended to deal with differential equations of arbitrary order, with any number of dependent and independent variables.

Suppose that a first-order ODE has a one-parameter Lie group of symmetries, whose tangent vector at (x, y) is (ξ, η). Then the partial differential operator

$$X = \xi(x, y)\partial_x + \eta(x, y)\partial_y \tag{2.73}$$

is called the *infinitesimal generator* of the Lie group. We have already encountered this operator; equations (2.28), which define canonical coordinates, can be rewritten as

$$Xr = 0, \qquad Xs = 1. \tag{2.74}$$

How is the infinitesimal generator affected by a change of coordinates? To find this out, suppose that (u, v) are the new coordinates and let $F(u, v)$ be an arbitrary smooth function. Then, by the chain rule,

$$\begin{aligned} XF(u, v) &= XF(u(x, y), v(x, y)) \\ &= \xi\{u_x F_u + v_x F_v\} + \eta\{u_y F_u + v_y F_v\} \\ &= (Xu)F_u + (Xv)F_v. \end{aligned}$$

However $F(u, v)$ is arbitrary and therefore, in terms of the new coordinates, the infinitesimal generator is

$$X = (Xu)\partial_u + (Xv)\partial_v. \tag{2.75}$$

In particular, if $(u, v) = (r, s)$, then (2.74) yields

$$X = (Xr)\partial_r + (Xs)\partial_s = \partial_s. \tag{2.76}$$

In canonical coordinates, the tangent vector is $(0, 1)$, and therefore (2.76) is consistent with our definition of the infinitesimal generator. In fact, X represents the tangent vector field in all coordinate systems. If we regard $\{\partial_x, \partial_y\}$ as a basis for the space of vector fields on the plane, X *is* the tangent vector at (x, y).

The infinitesimal generator provides a coordinate-independent way of characterizing the action of Lie symmetries on functions. Suppose that $G(r, s)$ is a smooth function, and let

$$F(x, y) = G(r(x, y), s(x, y)).$$

At any noninvariant point (x, y), the Lie symmetries map $F(x,y)$ to

$$F(\hat{x}, \hat{y}) = G(\hat{r}, \hat{s}) = G(r, s+\varepsilon).$$

Applying Taylor's theorem and (2.76), we obtain

$$F(\hat{x}, \hat{y}) = \sum_{j=0}^{\infty} \frac{\varepsilon^j}{j!} \frac{\partial^j G}{\partial s^j}(r, s) = \sum_{j=0}^{\infty} \frac{\varepsilon^j}{j!} X^j G(r, s).$$

We now revert to using (x, y) as coordinates, which gives

$$F(\hat{x}, \hat{y}) = \sum_{j=0}^{\infty} \frac{\varepsilon^j}{j!} X^j F(x, y). \tag{2.77}$$

If the expansion on the right-hand side of (2.77) converges, it is called the *Lie series* of F about (x, y). We have so far assumed that (x, y) is not an invariant point, but (2.77) is also valid at all invariant points. The reason for this is that $X = 0$ at any invariant point, so the Lie series has only the $j = 0$ term, which is $F(x, y)$.

A convenient shorthand for the Lie series (2.77) is

$$F(\hat{x}, \hat{y}) = e^{\varepsilon X} F(x, y); \tag{2.78}$$

this notation is suggested by the form of the Lie series. In particular, the Lie symmetries can be reconstructed as follows:

$$\begin{aligned} \hat{x} &= e^{\varepsilon X} x, \\ \hat{y} &= e^{\varepsilon X} y \end{aligned} \tag{2.79}$$

Therefore (2.78) amounts to the identity

$$F(e^{\varepsilon X} x, e^{\varepsilon X} y) = e^{\varepsilon X} F(x, y). \tag{2.80}$$

Everything in this section generalizes to any number of variables. Suppose that there are L variables, $z^1, \ldots, z^L$, and that the Lie symmetries are

$$\hat{z}^s(z^1, \ldots, z^L; \varepsilon) = z^s + \varepsilon \zeta^s(z^1, \ldots, z^L) + O(\varepsilon^2), \qquad s = 1, \ldots, L. \tag{2.81}$$

Then the infinitesimal generator of the one-parameter Lie group is

$$X = \zeta^s(z^1, \ldots, z^L) \frac{\partial}{\partial z^s}. \tag{2.82}$$

(The summation convention is used: if an index is used twice, one should sum over all possible values of that index.) Lie symmetries may be reconstructed from the Lie series:

$$\hat{z}^s = e^{\varepsilon X} z^s, \qquad s = 1, \dots, L. \tag{2.83}$$

More generally, if F is a smooth function,

$$F(e^{\varepsilon X} z^1, \dots, e^{\varepsilon X} z^L) = e^{\varepsilon X} F(z^1, \dots, z^L). \tag{2.84}$$

These general results will prove to be useful as we study higher-order differential equations.

Further Reading

Readers should consult Chapter 1 of Olver (1993) to find out more about the geometry of differential equations and symmetry groups. Ibragimov (1994) contains a classification of first-order ODEs that have various common Lie symmetries.

Exercises

2.1 Calculate the infinitesimal generator corresponding to each of the following one-parameter Lie groups of symmetries:
(a) $(\hat{x}, \hat{y}) = (x + \varepsilon,\ y + \varepsilon)$;
(b) $(\hat{x}, \hat{y}) = \left(\frac{x}{1-\varepsilon y}, \frac{y}{1-\varepsilon y}\right)$;
(c) $(\hat{x}, \hat{y}) = (x,\ e^{\varepsilon x} y)$.
Now find a pair of canonical coordinates for each generator.

2.2 Construct the one-parameter Lie groups corresponding to each of the following infinitesimal generators:
(a) $X = \partial_x + y\partial_y$;
(b) $X = (1 + x^2)\partial_x + xy\partial_y$;
(c) $X = 2xy\partial_x + (y^2 - x^2)\partial_y$.

2.3 Show that $(\hat{x}, \hat{y}) = (e^{\varepsilon} x,\ e^{\alpha\varepsilon} y)$ is a symmetry of $y' = 2y/x$ for every α and ε. Find every point that is invariant under each of these symmetries. For which α are the symmetries trivial?

2.4 Solve (2.60), using the symmetries that were derived in Example 2.13.

2.5 Show that $X = x\partial_x + 3y\partial_y$ generates Lie symmetries of the ODE

$$y' = \frac{3y}{x} + \frac{x^5}{2y + x^3}.$$

Use this result to solve the ODE.

2.6 The ODE

$$y' = e^{-x}y^2 + y + e^x$$

has a symmetry generator such that ξ and η are linear functions of x and y. Find this generator and use it to solve the ODE.

2.7 Repeat the last question for the ODE

$$y' = \frac{y + y^3}{x + (x+1)y^2}.$$

2.8 What is the most general first-order ODE that has the Lie symmetries $(\hat{x}, \hat{y}) = (e^{\varepsilon}x, e^{\alpha\varepsilon}y)$? (Treat α as a fixed constant.)

2.9 Suppose that $\bar{Q}_1$ and $\bar{Q}_2$ are linearly independent reduced characteristics of symmetries of the ODE $y' = \omega(x, y)$. Show that the general solution of the ODE is $\bar{Q}_1 = c\bar{Q}_2$, where c is an arbitrary constant.

2.10 Use the Lie series (2.77) to verify that (2.79) holds for each of the following infinitesimal generators:

(a) $X = x\partial_x - y\partial_y$;

(b) $X = x^2\partial_x + xy\partial_y$;

(c) $X = -y\partial_x + x\partial_y$.

3

How to Find Lie Point Symmetries of ODEs

Good hunting!

(Rudyard Kipling: The Jungle Book)

3.1 The Symmetry Condition

We have now met many of the basic ideas behind symmetry methods in the context of first-order ODEs. These ideas can be extended and applied to higher-order ODEs. For simplicity, we shall consider only ODEs of the form

$$y^{(n)} = \omega\big(x, y, y', \ldots, y^{(n-1)}\big), \qquad y^{(k)} \equiv \frac{d^k y}{dx^k}. \tag{3.1}$$

It is assumed that ω is (locally) a smooth function of all of its arguments.

We begin by stating the symmetry condition and examining some of its consequences. (The detailed justification of the symmetry condition is deferred until the end of this chapter.) A symmetry of (3.1) is a diffeomorphism that maps the set of solutions of the ODE to itself. Any diffeomorphism,

$$\Gamma : (x, y) \mapsto (\hat{x}, \hat{y}), \tag{3.2}$$

maps smooth planar curves to smooth planar curves. This action of Γ on the plane induces an action on the derivatives $y^{(k)}$, which is the mapping

$$\Gamma : \big(x, y, y', \ldots, y^{(n)}\big) \longmapsto \big(\hat{x}, \hat{y}, \hat{y}', \ldots, \hat{y}^{(n)}\big), \tag{3.3}$$

where

$$\hat{y}^{(k)} = \frac{d^k \hat{y}}{d\hat{x}^k}, \qquad k = 1, \ldots, n. \tag{3.4}$$

This mapping is called the nth *prolongation* of Γ. The functions $\hat{y}^{(k)}$ are calculated recursively (using the chain rule) as follows:

$$\hat{y}^{(k)} = \frac{d\hat{y}^{(k-1)}}{d\hat{x}} = \frac{D_x \hat{y}^{(k-1)}}{D_x \hat{x}}, \qquad \hat{y}^{(0)} \equiv \hat{y}. \tag{3.5}$$

Here D_x is the total derivative with respect to x:

$$D_x = \partial_x + y' \partial_y + y'' \partial_{y'} + \cdots. \tag{3.6}$$

The symmetry condition for the ODE (3.1) is

$$\hat{y}^{(n)} = \omega\big(\hat{x}, \hat{y}, \hat{y}', \dots, \hat{y}^{(n-1)}\big) \qquad \text{when (3.1) holds,} \tag{3.7}$$

where the functions $\hat{y}^{(k)}$ are given by (3.5).

For almost all ODEs, the symmetry condition (3.7) is nonlinear. Lie symmetries are obtained by linearizing (3.7) about $\varepsilon = 0$. No such linearization is possible for discrete symmetries, which makes them hard to find. However, it is usually easy to find out whether or not a given diffeomorphism is a symmetry of a particular ODE.

Example 3.1 Here we show that the transformation

$$(\hat{x}, \hat{y}) = \left(\frac{1}{x}, \frac{y}{x}\right) \tag{3.8}$$

is a symmetry of the second-order ODE

$$y'' = 0, \qquad x > 0. \tag{3.9}$$

From (3.5), we obtain

$$\hat{y}' = \frac{D_x(y/x)}{D_x(1/x)} = y - xy',$$

$$\hat{y}'' = \frac{D_x(y - xy')}{D_x(1/x)} = x^3 y''.$$

Therefore the symmetry condition,

$$\hat{y}'' = 0 \quad \text{when} \quad y'' = 0,$$

is satisfied. This symmetry is its own inverse, and so it belongs to a discrete group of order 2. The general solution of the ODE,

$$y = c_1 x + c_2, \tag{3.10}$$

is mapped by (3.8) to the solution

$$\hat{y} = \frac{y}{x} = c_1 + \frac{c_2}{x} = c_1 + c_2\hat{x}.$$

Hence this symmetry acts on the set of solution curves by exchanging the constants of integration c_1 and c_2.

The linearized symmetry condition for Lie symmetries is derived by the same method that we used for first-order ODEs. The trivial symmetry corresponding to $\varepsilon = 0$ leaves every point unchanged. Therefore, for ε sufficiently close to zero, the prolonged Lie symmetries are of the form

$$\begin{aligned} \hat{x} &= x + \varepsilon\xi + O(\varepsilon^2), \\ \hat{y} &= y + \varepsilon\eta + O(\varepsilon^2), \\ \hat{y}^{(k)} &= y^{(k)} + \varepsilon\eta^{(k)} + O(\varepsilon^2), \qquad k \geq 1. \end{aligned} \tag{3.11}$$

[N.B. The superscript in $\eta^{(k)}$ is merely an index; it does not denote a derivative of η.] We substitute (3.11) into the symmetry condition (3.7); the $O(\varepsilon)$ terms yield the linearized symmetry condition:

$$\eta^{(n)} = \xi\omega_x + \eta\omega_y + \eta^{(1)}\omega_{y'} + \cdots + \eta^{(n-1)}\omega_{y^{(n-1)}} \qquad \text{when (3.1) holds.} \tag{3.12}$$

The functions $\eta^{(k)}$ are calculated recursively from (3.5), as follows. For $k = 1$, we obtain

$$\hat{y}^{(1)} = \frac{D_x\hat{y}}{D_x\hat{x}} = \frac{y' + \varepsilon D_x\eta + O(\varepsilon^2)}{1 + \varepsilon D_x\xi + O(\varepsilon^2)} = y' + \varepsilon\big(D_x\eta - y'D_x\xi\big) + O(\varepsilon^2). \tag{3.13}$$

Therefore, from (3.11),

$$\eta^{(1)} = D_x\eta - y'D_x\xi. \tag{3.14}$$

Similarly,

$$\hat{y}^{(k)} = \frac{y^{(k)} + \varepsilon D_x\eta^{(k-1)} + O(\varepsilon^2)}{1 + \varepsilon D_x\xi + O(\varepsilon^2)},$$

and hence

$$\eta^{(k)}\big(x, y, y', \ldots, y^{(k)}\big) = D_x\eta^{(k-1)} - y^{(k)}D_x\xi. \tag{3.15}$$

The functions ξ, η and $\eta^{(k)}$ can all be written in terms of the characteristic, $Q = \eta - y'\xi$, as follows:

$$\begin{aligned} \xi &= -Q_{y'}, \\ \eta &= Q - y'Q_{y'}, \\ \eta^{(k)} &= D_x^k Q - y^{(k+1)} Q_{y'}, \qquad k \geq 1. \end{aligned} \tag{3.16}$$

[The derivation of (3.16) is left to the reader as an exercise.] This result is useful for computational purposes and readily generalizes to symmetries other than point symmetries (see Chapter 7).

For first-order ODEs, the right-hand side of the linearized symmetry condition (3.12) is $X\omega$, where X is the infinitesimal generator

$$X = \xi \partial_x + \eta \partial_y.$$

Recall that the infinitesimal generator is associated with the tangent vector to the orbit passing through (x, y), namely

$$(\xi, \eta) = \left(\frac{d\hat{x}}{d\varepsilon}, \frac{d\hat{y}}{d\varepsilon} \right) \bigg|_{\varepsilon=0}.$$

To deal with the action of Lie symmetries on derivatives of order n or smaller, we introduce the *prolonged infinitesimal generator*

$$X^{(n)} = \xi \partial_x + \eta \partial_y + \eta^{(1)} \partial_{y'} + \cdots + \eta^{(n)} \partial_{y^{(n)}}. \tag{3.17}$$

The coefficient of $\partial_{y^{(k)}}$ is the $O(\varepsilon)$ term in the expansion of $\hat{y}^{(k)}$, and so $X^{(n)}$ is associated with the tangent vector in the space of variables $(x, y, y', \ldots, y^{(n)})$. We can use the prolonged infinitesimal generator to write the linearized symmetry condition (3.12) in a compact form:

$$X^{(n)}\left(y^{(n)} - \omega\big(x, y, y', \ldots, y^{(n-1)}\big) \right) = 0 \qquad \text{when (3.1) holds.} \tag{3.18}$$

3.2 The Determining Equations for Lie Point Symmetries

Every symmetry that we have met so far is a diffeomorphism of the form

$$(\hat{x}, \hat{y}) = \big(\hat{x}(x, y),\ \hat{y}(x, y)\big). \tag{3.19}$$

This type of diffeomorphism is called a *point transformation*; any point transformation that is also a symmetry is called a *point symmetry*. For now, we restrict attention to point symmetries; other types of symmetries are discussed in Chapter 7.

To find the Lie point symmetries of an ODE (3.1), we must first calculate $\eta^{(k)}$, $k = 1, \dots, n$. The functions ξ and η depend upon x and y only, and therefore (3.14) and (3.15) give the following results.

$$\eta^{(1)} = \eta_x + (\eta_y - \xi_x)y' - \xi_y y'^2; \tag{3.20}$$

$$\begin{aligned}\eta^{(2)} &= \eta_{xx} + (2\eta_{xy} - \xi_{xx})y' + (\eta_{yy} - 2\xi_{xy})y'^2 - \xi_{yy}y'^3 \\ &\quad + \{\eta_y - 2\xi_x - 3\xi_y y'\}y'';\end{aligned} \tag{3.21}$$

$$\begin{aligned}\eta^{(3)} &= \eta_{xxx} + (3\eta_{xxy} - \xi_{xxx})y' + 3(\eta_{xyy} - \xi_{xxy})y'^2 + (\eta_{yyy} - 3\xi_{xyy})y'^3 \\ &\quad - \xi_{yyy}y'^4 + 3\{\eta_{xy} - \xi_{xx} + (\eta_{yy} - 3\xi_{xy})y' - 2\xi_{yy}y'^2\}y'' \\ &\quad - 3\xi_y y''^2 + \{\eta_y - 3\xi_x - 4\xi_y y'\}y'''.\end{aligned} \tag{3.22}$$

The number of terms in $\eta^{(k)}$ increases exponentially with k, so computer algebra is recommended for the study of high-order ODEs. However, the basic technique for finding Lie point symmetries can be learned by studying low-order ODEs. It is important to master this technique before resorting to computer algebra; practice is essential!

We begin by considering second-order ODEs of the form

$$y'' = \omega(x, y, y'). \tag{3.23}$$

The linearized symmetry condition is obtained by substituting (3.20) and (3.21) into (3.12) and then replacing y'' by $\omega(x, y, y')$. This gives

$$\begin{aligned}&\eta_{xx} + (2\eta_{xy} - \xi_{xx})y' + (\eta_{yy} - 2\xi_{xy})y'^2 - \xi_{yy}y'^3 + \{\eta_y - 2\xi_x - 3\xi_y y'\}\omega \\ &\quad = \xi\omega_x + \eta\omega_y + \{\eta_x + (\eta_y - \xi_x)y' - \xi_y y'^2\}\,\omega_{y'}.\end{aligned} \tag{3.24}$$

Although (3.24) looks complicated, it is commonly easy to solve. Both ξ and η are independent of y', and therefore (3.24) can be decomposed into a system of PDEs, which are the *determining equations* for the Lie point symmetries. The following examples illustrate the procedure.

Example 3.2 Consider the simplest second-order ODE,

$$y'' = 0. \tag{3.25}$$

The linearized symmetry condition for this ODE is

$$\eta^{(2)} = 0 \qquad \text{when} \quad y'' = 0,$$

that is,

$$\eta_{xx} + (2\eta_{xy} - \xi_{xx})y' + (\eta_{yy} - 2\xi_{xy})y'^2 - \xi_{yy}y'^3 = 0.$$

As ξ and η are independent of y', the linearized symmetry condition splits into the following system of determining equations:

$$\eta_{xx} = 0, \qquad 2\eta_{xy} - \xi_{xx} = 0, \qquad \eta_{yy} - 2\xi_{xy} = 0, \qquad \xi_{yy} = 0. \tag{3.26}$$

The general solution of the last of (3.26) is

$$\xi(x, y) = A(x)y + B(x),$$

for arbitrary functions A and B. The third of (3.26) gives

$$\eta(x, y) = A'(x)y^2 + C(x)y + D(x),$$

where C and D are also arbitrary functions. Then the remaining equations in (3.26) amount to

$$A'''(x)y^2 + C''(x)y + D''(x) = 0, \quad 3A''(x)y + 2C'(x) - B''(x) = 0. \tag{3.27}$$

Equating powers of y in (3.27), we obtain a system of ODEs for the unknown functions A, B, C, and D:

$$A''(x) = 0, \qquad C''(x) = 0, \qquad D''(x) = 0, \qquad B''(x) = 2C'(x).$$

These ODEs are easily solved, leading to the following result. For every one-parameter Lie group of symmetries of (3.25), the functions ξ and η are of the form

$$\xi(x, y) = c_1 + c_3 x + c_5 y + c_7 x^2 + c_8 xy,$$
$$\eta(x, y) = c_2 + c_4 y + c_6 x + c_7 xy + c_8 y^2,$$

where (as usual) $c_1, \ldots, c_8$ are constants. Therefore the most general infinitesimal generator is

$$X = \sum_{i=1}^{8} c_i X_i,$$

where

$$\begin{aligned} X_1 &= \partial_x, \quad X_2 = \partial_y, \quad X_3 = x\partial_x, \quad X_4 = y\partial_y, \quad X_5 = y\partial_x, \\ X_6 &= x\partial_y, \quad X_7 = x^2\partial_x + xy\partial_y, \quad X_8 = xy\partial_x + y^2\partial_y. \end{aligned} \tag{3.28}$$

Example 3.3 The ODE

$$y'' = \frac{y'^2}{y} - y^2 \tag{3.29}$$

arises in the study of swimming micro-organisms. The linearized symmetry condition is

$$\begin{aligned} &\eta_{xx} + (2\eta_{xy} - \xi_{xx})y' + (\eta_{yy} - 2\xi_{xy})y'^2 - \xi_{yy}y'^3 \\ &\quad + \{\eta_y - 2\xi_x - 3\xi_y y'\}\left(\frac{y'^2}{y} - y^2\right) \\ &\quad = \eta\left(-\frac{y'^2}{y^2} - 2y\right) + \{\eta_x + (\eta_y - \xi_x)y' - \xi_y y'^2\}\left(\frac{2y'}{y}\right). \end{aligned} \tag{3.30}$$

By comparing powers of y', we obtain the determining equations:

$$\begin{aligned} \xi_{yy} + \frac{1}{y}\xi_y &= 0, \\ \eta_{yy} - 2\xi_{xy} - \frac{1}{y}\eta_y + \frac{1}{y^2}\eta &= 0, \\ 2\eta_{xy} - \xi_{xx} + 3y^2\xi_y - \frac{2}{y}\eta_x &= 0, \\ \eta_{xx} - y^2(\eta_y - 2\xi_x) + 2y\eta &= 0. \end{aligned} \tag{3.31}$$

The first of (3.31) is readily integrated to give

$$\xi = A(x)\ln|y| + B(x); \tag{3.32}$$

then the second of (3.31) yields

$$\eta = A'(x)y\big(\ln|y|\big)^2 + C(x)y\ln|y| + D(x)y. \tag{3.33}$$

Here A, B, C, and D are unknown functions that are determined by the remaining equations of (3.31). Substituting (3.32) and (3.33) into the third determining equation, we obtain

$$3A''(x)\ln|y| + 3A(x)y + 2C'(x) - B''(x) = 0.$$

Therefore

$$A(x) = 0, \qquad B''(x) = 2C'(x). \tag{3.34}$$

Now we substitute (3.32) and (3.33) into the last of the determining equations (3.31), bearing in mind that $A = 0$. This leads to

$$C(x)y^2 \ln|y| + C''(x)y \ln|y| + \big(2B'(x) - C(x) + D(x)\big)y^2 + D''(x)y = 0,$$

which splits into the system

$$C(x) = 0, \qquad D(x) = -2B'(x), \qquad D''(x) = 0.$$

Taking (3.34) into account, we find that

$$B(x) = c_1 + c_2 x, \qquad D(x) = -c_2,$$

where c_1 and c_2 are arbitrary constants. Hence the general solution of the linearized symmetry condition is

$$\xi = c_1 + c_2 x, \qquad \eta = -2c_2 y. \tag{3.35}$$

Every infinitesimal generator is of the form

$$X = c_1 X_1 + c_2 X_2,$$

where

$$X_1 = \partial_x, \qquad X_2 = x\partial_x - 2y\partial_y. \tag{3.36}$$

Let $\mathcal{L}$ denote the set of all infinitesimal generators of one-parameter Lie groups of point symmetries of an ODE of order $n \geq 2$. The linearized symmetry condition is linear in ξ and η, and so

$$X_1,\ X_2 \in \mathcal{L} \Rightarrow c_1 X_1 + c_2 X_2 \in \mathcal{L}, \qquad \forall c_1,\ c_2 \in \mathbb{R}.$$

Therefore $\mathcal{L}$ is a vector space. The dimension, R, of this vector space is the number of arbitrary constants that appear in the general solution of the linearized symmetry condition. As in the above examples, every $X \in \mathcal{L}$ may be written in the form

$$\sum_{i=1}^{R} c_i X_i, \qquad c_i \in \mathbb{R}, \tag{3.37}$$

where $\{X_1, \ldots, X_R\}$ is a basis for $\mathcal{L}$. The set of point symmetries generated by all $X \in \mathcal{L}$ forms an R-parameter (local) Lie group. We shall call this "the group generated by $\mathcal{L}$" from here on.

The order of the ODE places restrictions upon R. For second-order ODEs, R is 0, 1, 2, 3, or 8. Moreover, R is 8 if and only if the ODE either is linear, or is linearizable by a point transformation. Every ODE of order $n \geq 3$ has $R \leq n+4$; if the ODE is linear (or linearizable), then $R \in \{n+1, n+2, n+4\}$. We shall not prove these results, but they provide us with a useful test. If one obtains a "general solution" of the linearized symmetry condition that contravenes these results, an error has been made.

For most second-order ODEs that arise from applications, ω is polynomial in y'. This makes it easy to split the linearized symmetry condition into the determining equations, by reading off all terms that are multiplied by a particular power of y'. For more general ω, splitting is achieved by collecting together all terms whose ratio is independent of y'.

The same technique works for higher-order ODEs. As a general rule, it is best to begin by calculating only those determining equations that are multiplied by the highest power of $y^{(n-1)}$. This gives some information about ξ and η. Then look at the next highest power of $y^{(n-1)}$ to find out more, and so on. The following example shows how this is done.

Example 3.4 The following ODE occurs in the study of flow in thin films with free boundaries:

$$y''' = y^{-3}. \tag{3.38}$$

The linearized symmetry condition for this ODE is

$$\eta^{(3)} = -3y^{-4}\eta \qquad \text{when (3.38) holds.} \tag{3.39}$$

From (3.22), we see that (3.39) is quadratic in y''. The terms involving y''^2 give

$$-3\xi_y = 0.$$

Therefore

$$\xi = A(x), \tag{3.40}$$

for some function A. Taking (3.40) into account, the terms in (3.39) that have a factor y'' yield the following:

$$\eta_{xy} - A''(x) + \eta_{yy}y' = 0.$$

Equating powers of y' in the above, we obtain

$$\eta = \big(A'(x) + c_1\big)y + B(x), \tag{3.41}$$

where B is a function and c_1 is constant. The remaining terms in the linearized symmetry condition reduce to

$$A''''(x)y + B'''(x) + 2A'''(x)y' + \big(c_1 - 2A'(x)\big)y^{-3}$$
$$= -3\big\{\big(A'(x) + c_1\big)y + B(x)\big\}y^{-4}.$$

Therefore

$$A = -4c_1 x + c_2, \qquad B = 0,$$

and so $\mathcal{L}$ is two dimensional, with a basis

$$X_1 = -4x\partial_x - 3y\partial_y, \qquad X_2 = \partial_x.$$

Later, we shall use the equivalent basis

$$X_1 = \partial_x, \qquad X_2 = x\partial_x + \tfrac{3}{4}y\partial_y, \tag{3.42}$$

to derive some exact solutions of (3.38).

In the above example, the fact that ω is independent of y'' immediately places severe restrictions upon ξ and η. This result generalizes, as follows. For any ODE of the form

$$y^{(n)} = \omega\big(x, y, y', \ldots, y^{(n-2)}\big), \qquad n \geq 3, \tag{3.43}$$

the linearized symmetry condition constrains ξ and η to be of the form

$$\xi = A(x), \qquad \eta = \big(\tfrac{1}{2}(n-1)A'(x) + c_1\big)y + B(x). \tag{3.44}$$

Here A and B are functions and c_1 is a constant.

3.3 Linear ODEs

Linear ODEs of order $n \geq 2$ prove to be surprisingly resistant to symmetry analysis, even though there is no shortage of Lie point symmetries. [Recall that for linear ODEs, $R = \dim(\mathcal{L}) \geq n+1$.] The problem is that most of the Lie point symmetries cannot be found until the general solution of the ODE is known.

To illustrate this, consider the second-order linear ODE

$$y'' = p(x)y' + q(x)y, \tag{3.45}$$

where $p(x)$ and $q(x)$ are given. The linearized symmetry condition leads to the result

$$\xi = A(x)y + B(x), \qquad \eta = \{A'(x) + p(x)A(x)\}y^2 + C(x)y + D(x),$$

where

$$\begin{aligned} A'' + (pA)' - qA &= 0, \\ B'' + (pB)' &= 2C', \\ C'' - pC' &= 2qB' + q'B, \\ D'' - pD' - qD &= 0. \end{aligned}$$

The equation for D is the original ODE, and A satisfies the *adjoint* ODE. Suppose that the general solution of (3.45) is

$$y = k_1 y_1(x) + k_2 y_2(x),$$

where k_1 and k_2 are arbitrary constants. Then the general solution of the system of ODEs for A, B, C, and D is

$$\begin{aligned} A &= e^{-\int p(x)dx}(c_4 y_1 + c_5 y_2), \\ B &= e^{-\int p(x)dx}\left(c_6 y_1^2 + 2c_7 y_1 y_2 + c_8 y_2^2\right), \\ C &= c_1 + e^{-\int p(x)dx}\left(c_6 y_1 y_1' + c_7(y_1' y_2 + y_1 y_2') + c_8 y_2 y_2'\right), \\ D &= c_2 y_1 + c_3 y_2. \end{aligned}$$

Hence the vector space of infinitesimal generators is eight dimensional and has a basis

$$\begin{aligned} X_1 &= y\partial_y, \\ X_2 &= y_1\partial_y, \\ X_3 &= y_2\partial_y, \\ X_4 &= e^{-\int p(x)dx}\left(y_1 y\partial_x + y_1' y^2\partial_y\right), \\ X_5 &= e^{-\int p(x)dx}\left(y_2 y\partial_x + y_2' y^2\partial_y\right), \\ X_6 &= e^{-\int p(x)dx}\left(y_1^2\partial_x + y_1 y_1' y\partial_y\right), \\ X_7 &= e^{-\int p(x)dx}\left(2y_1 y_2\partial_x + (y_1' y_2 + y_1 y_2')y\partial_y\right), \\ X_8 &= e^{-\int p(x)dx}\left(y_2^2\partial_x + y_2 y_2' y\partial_y\right). \end{aligned}$$

Every generator except X_1 depends upon the solutions of the original ODE, so it is not usually possible to solve the linearized symmetry condition completely.

Every homogeneous linear ODE of order $n \geq 3$ has infinitesimal generators of the form

$$X_1 = y\partial_y, \qquad X_2 = y_1\partial_y, \ \ldots, \ X_{n+1} = y_n\partial_y,$$

where $\{y_1, \ldots, y_n\}$ is a set of functionally independent solutions of the ODE. Some linear ODEs have one more infinitesimal generator. Other linear ODEs have three more and can be mapped to the ODE $y^{(n)} = 0$ by a point transformation. The three extra infinitesimal generators for

$$y^{(n)} = 0$$

are

$$X_{n+2} = \partial_x, \qquad X_{n+3} = x\partial_x, \qquad X_{n+4} = x^2\partial_x + (n-1)xy\partial_y.$$

3.4 Justification of the Symmetry Condition

At the start of this chapter, the symmetry condition was stated with little attempt at justification. Having seen how the linearized symmetry condition enables us to find Lie symmetries systematically, let us now think a little about the origin of the symmetry condition. Consider a diffeomorphism,

$$\Gamma : (x, y) \mapsto (\hat{x}, \hat{y}), \tag{3.46}$$

that maps a solution curve $y = f(x)$ to a "new" curve $\hat{y} = \tilde{f}(\hat{x})$. The function $f(x)$ satisfies

$$f^{(n)}(x) = \omega\big(x, f(x), f'(x), \ldots, f^{(n-1)}(x)\big), \tag{3.47}$$

because $y = f(x)$ is a solution of the ODE (3.1). Clearly, Γ is a symmetry only if the new curve is also a solution of the ODE, that is,

$$\tilde{f}^{(n)}(\hat{x}) = \omega\big(\hat{x}, \tilde{f}(\hat{x}), \tilde{f}'(\hat{x}), \ldots, \tilde{f}^{(n-1)}(\hat{x})\big). \tag{3.48}$$

We hardly ever know the general solution of the ODE in advance, so it is not practical to use (3.48) as a test. Nevertheless, any symmetry condition must tell us whether or not (3.48) is satisfied for each solution $y = f(x)$.

We achieve this by working in the $(n+2)$-dimensional Euclidean space of variables $(x, y, y', \ldots, y^{(n)})$, which is called the nth *jet space*, J^n. The ODE (3.1) defines a (hyper-) surface, $\mathcal{S}$, in J^n. For example, the ODE $y' = \omega(x, y)$ is represented as a surface in J^1, which is the space whose coordinates are (x, y, y').

Any smooth curve $y = f(x)$ in the plane is represented in J^n by the unique smooth curve

$$\big(x, y, y', \ldots, y^{(n)}\big) = \big(x, f(x), f'(x), \ldots, f^{(n)}(x)\big). \tag{3.49}$$

The curve (3.49) is called the *lift* of $y = f(x)$; the term *lifted curve* is also used. The (x, y) coordinates of any lift determine the original curve on the plane. This correspondence between smooth curves on the plane and their lifts means that any smooth mapping (3.46) between curves on the plane can be extended to J^n:

$$\Gamma : \left(x, y, y', \ldots, y^{(n)}\right) \longmapsto \left(\hat{x}, \hat{y}, \hat{y}', \ldots, \hat{y}^{(n)}\right), \tag{3.50}$$

where, by definition of the lift,

$$\hat{y}^{(k)} = \frac{d^k \hat{y}}{d\hat{x}^k}, \qquad k = 1, \ldots, n. \tag{3.51}$$

This extension of the action of Γ to all derivatives of order n or less is the nth prolongation of Γ.

From (3.47), if $y = f(x)$ is a solution of the ODE then its lift (3.49) lies within $\mathcal{S}$. Conversely, any lifted curve that lies within $\mathcal{S}$ is a solution of the ODE. The nth prolongation of Γ maps $\mathcal{S}$ to a new surface, $\tilde{\mathcal{S}}$, which is of the form

$$\hat{y}^{(n)} = \tilde{\omega}\left(\hat{x}, \hat{y}, \hat{y}', \ldots, \hat{y}^{(n-1)}\right). \tag{3.52}$$

In particular, if Γ maps a solution $y = f(x)$ to the curve $\hat{y} = \tilde{f}(\hat{x})$, then the lift of the new curve lies within $\tilde{\mathcal{S}}$, that is,

$$\tilde{f}^{(n)}(\hat{x}) = \tilde{\omega}\left(\hat{x}, \tilde{f}(\hat{x}), \tilde{f}'(\hat{x}), \ldots, \tilde{f}^{(n-1)}(\hat{x})\right). \tag{3.53}$$

So (3.48) holds if and only if $\tilde{\omega} = \omega$ on the lift of $y = f(x)$. Until now, we have considered a single solution curve. The equivalent condition for the whole set of solution curves is that $\tilde{\omega} = \omega$ on $\mathcal{S}$, that is,

$$\hat{y}^{(n)} = \omega\left(\hat{x}, \hat{y}, \hat{y}', \ldots, \hat{y}^{(n-1)}\right) \quad \text{when (3.1) holds,} \tag{3.54}$$

Therefore $\mathcal{S}$ is mapped to itself by the (prolonged) action of any symmetry. In other words, $\mathcal{S}$ is invariant under any symmetry. Of course, this is to be expected from our premise that a symmetry is a diffeomorphism that leaves an object apparently unchanged.

N.B. The above discussion shows that some care is needed in defining symmetries. Not every diffeomorphism mapping the surface $\mathcal{S}$ to itself is a symmetry. There is an extra condition (3.51), which ensures that the diffeomorphism maps each lift to a lift (at least, for those lifts that lie in $\mathcal{S}$). This condition is the requirement that symmetries preserve the "structure" of solutions of differential equations.

Notes and Further Reading

In principle, the symmetry condition could be used to find the discrete point symmetries of a given ODE of order $n \geq 2$. The functions $\hat{y}^{(k)}$ are determined by the prolongation formulae (3.5). Although the symmetry condition is a highly nonlinear partial differential equation, it can be simplified somewhat because $\hat{x}$ and $\hat{y}$ are independent of $y', y'', \ldots, y^{(n-1)}$. This splits the symmetry condition into a system of coupled nonlinear PDEs. However, such systems are generally extremely difficult to solve. A much easier method of obtaining the discrete symmetries is described in Chapter 11.

Lie algebras of symmetry generators for ODEs are subject to various constraints (some of which are mentioned in §3.2). There are enough constraints for a complete classification to be made of all Lie algebras of point symmetry generators. Olver (1995) contains further details and references.

The result (3.44) simplifies the calculation of symmetry generators for a large class of ODEs. Many similar labour-saving results are described in Bluman and Kumei (1989).

The justification of the symmetry condition in §3.4 is based on the action of a symmetry on an individual solution curve. In particular, the lift is used to derive the prolongation formulae. Some authors derive the same formulae using the idea of tangential contact between pairs of curves. For a comparison of these two approaches, see Sewell and Roulstone (1994).

Exercises

3.1 Show that ξ, η and $\eta^{(k)}$ satisfy (3.16).

3.2 Derive (3.21) and (3.22) from the prolongation formula (3.15) and write down $\eta^{(4)}$ explicitly.

3.3 Calculate $X^{(4)}$ for each of the following generators:
- (a) $X = \partial_y$;
- (b) $X = x\partial_x + \alpha y\partial_y$, where α is constant;
- (c) $X = xy\partial_x + y^2\partial_y$;
- (d) $X = -y\partial_x + x\partial_y$.

3.4 Write down and solve the determining equations for the symmetry generators of $y'' = y'^4 + \alpha y'^2$ (where α is constant). What is the dimension of $\mathcal{L}$ for most values of α? For which α is the dimension of $\mathcal{L}$ greater than this?

3.5 Find the most general infinitesimal generator for the ODE

$$y''' = 7y' - 6y.$$

Note that $\dim(\mathcal{L}) < 7$ in this example, even though the ODE is linear.

3.6 The overdetermined set of PDEs for ξ and η are not always easy to solve. For example, if an ODE has a discrete symmetry $(\hat{x}, \hat{y}) = (y, x)$ then $\eta(x, y)$ and $\xi(y, x)$ satisfy the same PDEs, which are typically highly coupled. Write down the determining equations for the ODE $y'' = (1 - y')^3$. Now calculate the general solution of the determining equations. (Hint: Look for a suitable change of variables.)

3.7 It is sometimes useful to classify members of a family of ODEs according to their symmetries. Consider the ODEs of the form

$$y'' = f(y), \qquad f''(y) \neq 0.$$

Every such ODE has symmetries generated by $X_1 = \partial_x$. Use the linearized symmetry condition to find all functions $f(y)$ for which there are additional Lie point symmetries. [This is an example of a *group classification* problem. See Bluman and Kumei (1989) for further examples.]

4

How to Use a One-Parameter Lie Group

You know my methods.
Apply them.

(Sir Arthur Conan Doyle: A Study in Scarlet)

4.1 Reduction of Order by Using Canonical Coordinates

Now that we are able to find Lie symmetries systematically, how should they be used? To solve a first-order ODE, we write it in terms of canonical coordinates. Higher-order ODEs also benefit from canonical coordinates. Henceforth, differentiation with respect to r is denoted by a dot $(\cdot)$; for example, $\dot{s}$ denotes $\frac{ds}{dr}$.

Suppose that X is an infinitesimal generator of a one-parameter Lie group of symmetries of the ODE

$$y^{(n)} = \omega\big(x, y, y', \dots, y^{(n-1)}\big), \qquad n \geq 2. \tag{4.1}$$

Let (r, s) be canonical coordinates for the group generated by X, so that

$$X = \partial_s. \tag{4.2}$$

If the ODE (4.1) is written in terms of canonical coordinates, it is of the form

$$s^{(n)} = \Omega\big(r, s, \dot{s}, \dots, s^{(n-1)}\big), \qquad s^{(k)} = \frac{d^k s}{dr^k}. \tag{4.3}$$

for some Ω. However, the ODE (4.3) is invariant under the Lie group of translations in s, so the symmetry condition gives

$$\Omega_s = 0.$$

Therefore

$$s^{(n)} = \Omega\big(r, \dot{s}, \dots, s^{(n-1)}\big). \tag{4.4}$$

By writing the ODE (4.1) in terms of canonical coordinates, we have reduced it to an ODE of order $n - 1$ for $v = \dot{s}$:

$$v^{(n-1)} = \Omega\big(r, v, \ldots, v^{(n-2)}\big), \qquad v^{(k)} = \frac{d^{k+1}s}{dr^{k+1}}. \tag{4.5}$$

Suppose that the reduced ODE has the general solution

$$v = f(r; c_1, \ldots, c_{n-1}).$$

Then the general solution of the original ODE (4.1) is

$$s(x, y) = \int^{r(x,y)} f(r; c_1, \ldots, c_{n-1})\, dr + c_n.$$

More generally, if v is any function of $\dot{s}$ and r such that $v_{\dot{s}}(r, \dot{s}) \neq 0$, the ODE (4.4) reduces to an ODE of the form

$$v^{(n-1)} = \tilde{\Omega}\big(r, v, \ldots, v^{(n-2)}\big), \qquad v^{(k)} = \frac{d^k v}{dr^k}. \tag{4.6}$$

Once the general solution of (4.6) has been found, the relationship

$$\dot{s} = \dot{s}(r, v)$$

gives the general solution of (4.1):

$$s(x, y) = \int^{r(x,y)} \dot{s}\big(r,\ v(r; c_1, \ldots, c_{n-1})\big)\, dr + c_n. \tag{4.7}$$

To summarize: if we know a one-parameter Lie group of symmetries, we are able to solve (4.1) by solving a lower-order ODE, then integrating. This is encouraging, for we already have a systematic technique for deriving Lie point symmetries.

Example 4.1 Consider the linear ODE

$$y'' = \left(\frac{3}{x} - 2x\right)y' + 4y. \tag{4.8}$$

We do not yet know any solutions of (4.8), and so only one infinitesimal generator is available, namely

$$X = y\partial_y.$$

The simplest canonical coordinates are

$$r = x, \qquad s = \ln|y|,$$

which prolong to

$$\frac{ds}{dr} = \frac{y'}{y}, \qquad \frac{d^2s}{dr^2} = \frac{y''}{y} - \frac{y'^2}{y^2}.$$

Hence the ODE (4.8) reduces to the Riccati equation

$$\frac{dv}{dr} = \left(\frac{3}{r} - 2r\right)v + 4 - v^2, \qquad v = \frac{ds}{dr} = \frac{y'}{y}. \tag{4.9}$$

This reduction is a standard technique, which yields the general solution of the original ODE if one solution of the reduced equation can be found. The Riccati equation (4.9) has one simple solution:

$$v = -2r. \tag{4.10}$$

Then the remaining solutions are of the form

$$v = -2r + w^{-1}, \tag{4.11}$$

where w satisfies the linear ODE

$$\frac{dw}{dr} = 1 - \left(\frac{3}{r} + 2r\right)w.$$

Therefore

$$w = \frac{c_1}{r^3}e^{-r^2} + \frac{1}{2r} - \frac{1}{2r^3}. \tag{4.12}$$

Having found $\dot{s} = v(r; c_1)$, all that remains is to carry out the quadrature. The exceptional solution (4.10) gives

$$\ln|y| = s = -r^2 + c_2 = -x^2 + c_2.$$

Similarly, the general solution (4.11) yields

$$\ln|y| = \ln\left|c_1 e^{-x^2} + \tfrac{1}{2}(x^2 - 1)\right| + c_2.$$

Combining these results and redefining the arbitrary constants, we find that the general solution of (4.8) is

$$y = \tilde{c}_1 e^{-x^2} + \tilde{c}_2(x^2 - 1). \tag{4.13}$$

In the above example, it was convenient to take $v = \dot{s}$. Commonly, this is not the most convenient choice for v, as the following example shows.

Example 4.2 The nonlinear second-order ODE

$$y'' = \frac{y'^2}{y} + \left(y - \frac{1}{y}\right) y' \tag{4.14}$$

has a one-parameter Lie group of symmetries whose infinitesimal generator is

$$X = \partial_x.$$

These translations are the only Lie point symmetries of (4.14). The simplest canonical coordinates are obvious:

$$(r, s) = (y, x).$$

If we choose v to be $\dot{s} = (y')^{-1}$, the ODE (4.14) reduces to

$$\frac{dv}{dr} = -\frac{y''}{y'^3} = -\frac{v}{r} + \left(\frac{1}{r} - r\right) v^2.$$

This is a Bernoulli equation, which can be solved by rewriting it as a linear equation for v^{-1}.

Clearly, it is much easier to choose

$$v = y', \qquad \text{i.e.,} \ v = (\dot{s})^{-1},$$

from the outset. With this choice of v, the ODE (4.14) reduces directly to the linear ODE

$$\frac{dv}{dr} = \frac{y''}{y'} = \frac{v}{r} + r - \frac{1}{r}. \tag{4.15}$$

The general solution of (4.15) is

$$v = r^2 - 2c_1 r + 1.$$

(The factor -2 in front of the arbitrary constant c_1 is there for convenience.) Therefore

$$s = \int \frac{1}{v(r)} = \int \frac{dr}{r^2 - 2c_1 r + 1}.$$

After carrying out the (easy) quadrature, we obtain the general solution of (4.14):

$$y = \begin{cases} c_1 - \sqrt{c_1^2 - 1}\, \tanh\left(\sqrt{c_1^2 - 1}\,(x + c_2)\right), & c_1^2 > 1; \\ c_1 - (x + c_2)^{-1}, & c_1^2 = 1; \\ c_1 + \sqrt{1 - c_1^2}\, \tan\left(\sqrt{1 - c_1^2}\,(x + c_2)\right), & c_1^2 < 1. \end{cases} \tag{4.16}$$

In each of the previous examples, we were able to solve the reduced ODE to find $\dot{s}$ as a function of r. This is unusual; knowing a one-parameter group of symmetries enables us to reduce the order of the ODE, but there is no guarantee that the reduced ODE is easy to solve! However, suppose that we know more than one infinitesimal generator, that is, $R = \dim(\mathcal{L}) \geq 2$. Could these Lie symmetries be used to reduce the order of the ODE by R, or to solve it completely if $n \leq R$? To answer this question, we need to know more about the symmetries, as the next example shows.

Example 4.3 Consider the ODE

$$y''' = \frac{1}{y^3}, \qquad x > 0, \tag{4.17}$$

whose Lie point symmetries are generated by

$$X_1 = \partial_x, \qquad X_2 = x\partial_x + \tfrac{3}{4}y\partial_y.$$

The group generated by X_1 consists of translations, whereas X_2 generates scaling symmetries. First we reduce the ODE (4.17) by using X_1; the canonical coordinates $(r_1, s_1) = (y, x)$ yield

$$\frac{d^2v_1}{dr_1^2} = \frac{1}{r_1^3 v_1^2} - \frac{1}{v_1}\left(\frac{dv_1}{dr_1}\right)^2, \quad \text{where} \quad v_1 = (\dot{s})^{-1} = y'. \tag{4.18}$$

The only Lie point symmetries of the reduced ODE (4.18) are the scalings generated by

$$\tilde{X}_2 = \tfrac{3}{4}r_1\partial_{r_1} - \tfrac{1}{4}v_1\partial_{v_1}. \tag{4.19}$$

These symmetries are inherited from the original ODE (4.17); they arise from the unused one-parameter Lie group generated by X_2. To find out how this group acts on the reduced set of variables $(r_1, v_1) = (y, y')$, we need to prolong X_2:

$$X_2^{(1)} = x\partial_x + \tfrac{3}{4}y\partial_y - \tfrac{1}{4}y'\partial_{y'}.$$

The reduced generator $\tilde{X}_2$ is obtained by restricting attention to those terms in $X_2^{(1)}$ that act on (r_1, v_1). In this example, $\tilde{X}_2$ is independent of $s_1 = x$, so it generates point symmetries of the reduced ODE. These symmetries enable us to reduce the order again. If we use

$$r_2 = r_1^{\frac{1}{3}} v_1, \qquad v_2 = r_1^{\frac{4}{3}} \frac{dv_1}{dr_1},$$

the reduced ODE is

$$\frac{dv_2}{dr_2} = \frac{3 - 4r_2^2 v_2 - 3r_2 v_2^2}{r_2^2(r_2 + 3v_2)}. \tag{4.20}$$

The symmetries of (4.20) are not known, so we cannot proceed further. Nevertheless, we have succeeded in using both Lie symmetries of (4.17) to reduce the order by 2.

Surprisingly, the key to this success is that we used X_1 first. If any other infinitesimal generator is used first, the resulting second-order ODE does not inherit the unused symmetries. For example, if X_2 is used first, the ODE (4.17) reduces to

$$\frac{d^2v}{dr^2} = \frac{16 - 5r^3 v}{r^3(3r - 4v)^2} + \frac{4\left(\frac{dv}{dr}\right)^2 - 9\frac{dv}{dr}}{3r - 4v}, \tag{4.21}$$

where

$$r = x^{-\frac{3}{4}} y, \qquad v = x^{\frac{1}{4}} y'.$$

The reduced ODE (4.21) has no Lie point symmetries.

In the next chapter, we shall investigate the set of infinitesimal generators more thoroughly. This will enable us to find out which symmetries should be used first, if other symmetries are to be inherited by the reduced ODE.

4.2 Variational Symmetries

If an ODE is derived from a variational principle, it may be possible to use a one-parameter Lie group of point symmetries to reduce the order by 2. To begin with, consider the variational principle

$$\delta W = 0; \tag{4.22}$$

here the action, W, is of the form

$$W = \int L(x, y, y')\, dx, \tag{4.23}$$

where $L(x, y, y')$ is the Lagrangian. This variational principle leads to the Euler–Lagrange equation

$$L_y - D_x(L_{y'}) = 0. \tag{4.24}$$

Suppose that a point transformation maps W to

$$\hat{W} = \int L(\hat{x}, \hat{y}, \hat{y}')\, d\hat{x}, \tag{4.25}$$

where

$$\hat{W} = W. \tag{4.26}$$

Such a transformation is called a *variational symmetry.* The Euler–Lagrange equation for the transformed problem is

$$\hat{L}_{\hat{y}} - D_{\hat{x}}\left(\hat{L}_{\hat{y}'}\right) = 0, \qquad \text{where} \quad \hat{L} = L(\hat{x}, \hat{y}, \hat{y}'),$$

and so the variational symmetry is also a point symmetry of the Euler–Lagrange equation. If we can find the Lie point symmetries of the Euler–Lagrange equation, it is usually easy to check which of these are also variational symmetries.

Suppose that $X = \xi\partial_x + \eta\partial_y$ generates a one-parameter Lie group of point symmetries of the Euler–Lagrange equation (4.24). Then $\hat{W}$ may be expanded in powers of ε:

$$\hat{W} = \int \{L(x, y, y') + \varepsilon X^{(1)} L + O(\varepsilon^2)\}\{1 + \varepsilon(D_x\xi) + O(\varepsilon^2)\}\, dx.$$

Collecting together the first-order terms, $\hat{W} = W$ if

$$X^{(1)} L + (D_x\xi) L = 0. \tag{4.27}$$

In particular, $X = \partial_y$ generates variational symmetries if $L_y = 0$. One consequence of this condition is that the Euler–Lagrange equation is

$$D_x(L_{y'}) = 0,$$

and hence

$$L_{y'}(x, y') = c_1. \tag{4.28}$$

The reduced ODE (4.28) retains the Lie symmetries generated by $X = \partial_y$, so a further reduction of order is possible. This is easily done, by using (4.28) to write

$$y' = F(x; c_1)$$

(for some function F), and then integrating:

$$y = \int F(x; c_1)\, dx + c_2.$$

The above technique can also be used for other variational Lie point symmetries. Having ascertained that a particular generator X satisfies (4.27), one should rewrite the variational problem in terms of canonical coordinates. Then $X = \partial_s$ generates variational symmetries of

$$\delta W = \delta \int \tilde{L}(r, s, \dot{s})\, dr = 0, \tag{4.29}$$

where the change of variables in the integral requires that

$$\tilde{L}(r, s, \dot{s}) = \frac{L}{D_x r}. \tag{4.30}$$

The condition (4.27), rewritten in terms of canonical coordinates, amounts to $\tilde{L}_s = 0$. Therefore the Euler–Lagrange equation,

$$\tilde{L}_s - D_r(\tilde{L}_{\dot{s}}) = 0,$$

reduces to

$$\tilde{L}_{\dot{s}}(r, \dot{s}) = c_1. \tag{4.31}$$

The solution is completed by using (4.31) to write $\dot{s}$ as a function of r and c_1 then integrating to find s.

Example 4.4 One of the fundamental problems of classical mechanics is to compute the motion of an object described by a conservative system. Hamilton's principle states that the Lagrangian is

$$L = T - U,$$

where T is the object's kinetic energy and U is the potential energy. The simplest such systems are of the form

$$L = \tfrac{1}{2}y'^2 - U(y); \tag{4.32}$$

here $y(x)$ is the position of the object at time x. The Euler–Lagrange equation is

$$y'' = -G(y), \qquad \text{where} \quad G(y) = \frac{dU}{dy}. \tag{4.33}$$

Every ODE (4.33) has Lie symmetries generated by $X = \partial_x$; indeed, for most functions G, these are the only Lie point symmetries. These symmetries are variational, because

$$X^{(1)}L + (D_x\xi)L = L_x = 0.$$

Using the canonical coordinates $(r, s) = (y, x)$, the variational problem is equivalent to (4.29), where

$$\tilde{L}(r, \dot{s}) = \frac{L}{D_x r} = \frac{y'}{2} - \frac{U(y)}{y'} = \frac{1}{2\dot{s}} - U(r)\dot{s}.$$

Applying (4.31), we obtain the first integral

$$\tilde{L}_{\dot{s}} = -\frac{1}{2\dot{s}^2} - U(r) = c_1.$$

Reverting to the original variables,

$$\tfrac{1}{2}y'^2 + U(y) = -c_1, \tag{4.34}$$

and therefore

$$x = \pm\int \frac{dy}{\sqrt{-2(U(y) + c_1)}} + c_2. \tag{4.35}$$

ODEs of the form (4.33) are usually solved by the standard trick of multiplying both sides by y' and then integrating to obtain the separable ODE (4.34). This trick is nothing more than the exploitation of one particular Lie group of variational symmetries. For other classes of variational problem, the solution method may not be so obvious. Nevertheless, if variational symmetries exist, they can be found and used systematically.

Example 4.5 Consider the ODE

$$y'' = \frac{y'}{x} + \frac{3y^2}{2x^3}, \tag{4.36}$$

which arises from the variational problem whose Lagrangian is

$$L(x, y, y') = \frac{y'^2}{2x} + \frac{y^3}{2x^4}. \tag{4.37}$$

This ODE has the scaling symmetries generated by

$$X = x\partial_x + y\partial_y;$$

it has no other Lie point symmetries. However the symmetries generated by X are variational, because

$$X^{(1)}L + (D_x\xi)L = xL_x + yL_y + L = 0.$$

Using the canonical coordinates

$$(r, s) = \left(\frac{y}{x}, \ln|x|\right),$$

the transformed Lagrangian is

$$\tilde{L} = \frac{L}{D_x r} = \frac{y'^2 + \left(\frac{y}{x}\right)^3}{2\left(y' - \frac{y}{x}\right)} = \frac{1}{2\dot{s}} + r + \tfrac{1}{2}(r^3 + r^2)\dot{s}.$$

Therefore the Euler–Lagrange equation reduces to

$$-\frac{1}{2\dot{s}^2} + \tfrac{1}{2}(r^3 + r^2) = c_1,$$

and so the general solution is

$$\ln|x| = \pm \int^{\frac{y}{x}} \frac{dr}{\sqrt{r^3 + r^2 - 2c_1}}. \tag{4.38}$$

The same ideas carry over to higher-order variational problems. Given a Lagrangian

$$L = L\left(x, y, y', \ldots, y^{(p)}\right), \tag{4.39}$$

the Euler–Lagrange equation is

$$L_y - D_x(L_{y'}) + D_x^2(L_{y''}) - \cdots + (-1)^p D_x^p(L_{y^{(p)}}) = 0. \tag{4.40}$$

By the same argument as for $p = 1$, Lie point symmetries of the Euler–Lagrange equation are variational symmetries if

$$X^{(p)}L + (D_x\xi)L = 0. \tag{4.41}$$

Suppose that $X = \partial_y$ (if necessary, rewrite the problem in canonical coordinates first). Then (4.41) amounts to $L_y = 0$, and therefore the Euler–Lagrange equation reduces to

$$L_{y'} - D_x(L_{y''}) + \cdots + (-1)^{p-1} D_x^{p-1}(L_{y^{(p)}}) = c_1. \tag{4.42}$$

This reduced ODE has Lie point symmetries generated by X, and so it can be reduced further to

$$\bar{L}_v - D_r(\bar{L}_{\dot{v}}) + \cdots + (-1)^{p-1} D_r^{p-1}(\bar{L}_{v^{(p-1)}}) = 0, \tag{4.43}$$

where

$$(r, v) = (x, y'),$$
$$\bar{L}\left(r, v, \ldots, v^{(p-1)}\right) = L - c_1 v,$$

and

$$D_r = \partial_r + \dot{v}\partial_v + \ddot{v}\partial_{\dot{v}} + \cdots.$$

Clearly, (4.43) is the Euler–Lagrange equation for the reduced variational problem

$$\delta \int \bar{L}\, dr = 0.$$

If one can solve this problem (possibly with the aid of variational symmetries) then the solution of the original problem is obtained by quadrature:

$$y = \int^x v(r; c_1, \ldots, c_{2p-1})\, dr + c_{2p}.$$

4.3 Invariant Solutions

Many ODEs cannot be completely solved using their Lie point symmetries. Even so, it may be possible to derive solutions that are invariant under the group generated by a particular X. From (2.15), every curve C on the (x, y) plane that is invariant under the group generated by X satisfies

$$Q(x, y, y') = \eta - y'\xi = 0 \tag{4.44}$$

on C. All that is needed is to solve the first-order ODE (4.44), then to check which (if any) of these solutions satisfy the given ODE.

Example 4.6 The Blasius equation,

$$y''' = -yy'', \tag{4.45}$$

has translational and scaling symmetries generated by

$$X_1 = \partial_x, \qquad X_2 = x\partial_x - y\partial_y. \tag{4.46}$$

These symmetries enable the Blasius equation to be reduced to a first-order ODE whose solution is not known. So our only chance of finding exact solutions is to look for invariant solutions. For $X = X_1$, the invariant curve condition (4.44) reduces to $y' = 0$. Therefore every curve that is invariant under X_1 is of the form

$$y = c. \tag{4.47}$$

All such curves are solutions of the Blasius equation.

Now we seek solutions that are invariant under $X = X_2$. The invariant curve condition is

$$Q = -y - xy' = 0.$$

Therefore the invariant curves are

$$y = \frac{c}{x}, \qquad c \in \mathbb{R}.$$

The Blasius equation is satisfied if

$$y = 0 \quad \text{or} \quad y = \frac{3}{x} \tag{4.48}$$

The curve $y = 0$ is the only solution that is invariant under both X_1 and X_2.

Are there any other invariant solutions? We have not yet considered all possible one-parameter groups. This is easily done, because every remaining one-parameter Lie group is generated by $X = kX_1 + X_2$, for some nonzero k. The invariant curve condition is

$$Q = -y - (x + k)y' = 0,$$

which leads to the invariant solutions

$$y = 0, \qquad y = \frac{3}{x + k}. \tag{4.49}$$

These solutions can also be obtained by considering the action of the group generated by X_1 upon each of the invariant solutions (4.48). The symmetries generated by X_1 are

$$(\hat{x}, \hat{y}) = (x + \varepsilon, y), \qquad \varepsilon \in \mathbb{R}. \tag{4.50}$$

Consider the action of (4.50) upon the invariant solution $y = 3/x$. This solution may be rewritten as

$$\hat{y} = \frac{3}{\hat{x} - \varepsilon}; \tag{4.51}$$

it is invariant under

$$\begin{aligned} X_2 &= (X_2\hat{x})\partial_{\hat{x}} + (X_2\hat{y})\partial_{\hat{y}} \\ &= x\partial_{\hat{x}} - y\partial_{\hat{y}} \\ &= (\hat{x} - \varepsilon)\partial_{\hat{x}} - \hat{y}\partial_{\hat{y}}. \end{aligned}$$

We now introduce a useful notation; if

$$X_i = \xi_i(x, y)\partial_x + \eta_i(x, y)\partial_y$$

then let

$$\hat{X}_i = \xi_i(\hat{x}, \hat{y})\partial_{\hat{x}} + \eta_i(\hat{x}, \hat{y})\partial_{\hat{y}}. \tag{4.52}$$

Therefore (4.51) is invariant under

$$X_2 = \hat{X}_2 - \varepsilon\hat{X}_1,$$

Dropping carets, $y = 3/x$ is mapped to

$$y = \frac{3}{x - \varepsilon}, \tag{4.53}$$

which is invariant under $X_2 - \varepsilon X_1$. A similar calculation shows that $y = 0$ is mapped to itself by the action of the group generated by X_1.

The invariant canonical coordinate $r(x, y)$ satisfies

$$\xi D_x r + Q r_y = \xi r_x + \eta r_y = 0,$$

so every invariant solution on which $\xi \neq 0$ is of the form $r(x, y) = c$. There may also be invariant solutions $y = f(x)$ such that

$$\xi(x, f(x)) = \eta(x, f(x)) = 0 \tag{4.54}$$

Generally speaking, these are easily obtained by solving either $\xi(x, y) = 0$ or $\eta(x, y) = 0$, then checking that the solution satisfies the given ODE and (4.54).

There is another way to find invariant solutions on which ξ does not vanish, which is particularly useful if (4.44) is hard to solve. For Lie point symmetries, ξ and η are functions of x and y only. Therefore (4.44) holds if

$$y' = \frac{\eta(x, y)}{\xi(x, y)} \tag{4.55}$$

on invariant curves for which $\xi(x, y) \neq 0$. Higher derivatives are calculated by the usual prolongation formula. When the results are substituted into the ODE, one obtains an algebraic equation that defines all invariant solution curves. The next example illustrates the general procedure.

Example 4.7 Recall that the ODE

$$y''' = \frac{1}{y^3}, \qquad x > 0, \tag{4.56}$$

has $\mathcal{L}$ spanned by

$$X_1 = \partial_x, \qquad X_2 = x\partial_x + \tfrac{3}{4}y\partial_y.$$

Unlike the previous example, there are no solutions that are invariant under the group generated by X_1. The invariant curve condition for $X = X_2$ is

$$Q = \tfrac{3}{4}y - xy' = 0.$$

So, on every invariant curve,

$$y' = \frac{3y}{4x}. \tag{4.57}$$

Now we use (4.57) to calculate y'' and y''' on the invariant curves. First differentiate (4.57) with respect to x:

$$y'' = \frac{3y'}{4x} - \frac{3y}{4x^2}.$$

Then use (4.57) to determine y'' as a function of x and y:

$$y'' = \frac{9y}{16x^2} - \frac{3y}{4x^2} = -\frac{3y}{16x^2}.$$

Similarly,

$$y''' = -\frac{3y'}{16x^2} + \frac{3y}{8x^3} = \frac{15y}{64x^3} \tag{4.58}$$

on every invariant curve. Comparing (4.58) with (4.56), we see that the invariant solution curves are

$$y = \pm\left(\tfrac{64}{15}\right)^{\frac{1}{4}} x^{\frac{3}{4}}. \tag{4.59}$$

As in the previous example, there is a whole family of invariant solutions arising from the action of the group generated by X_1 on the solutions that are invariant under X_2. Specifically,

$$y = \pm\left(\tfrac{64}{15}\right)^{\frac{1}{4}} (x - \varepsilon)^{\frac{3}{4}}. \tag{4.60}$$

is invariant under $X_2 - \varepsilon X_1$.

Further Reading

Bluman and Kumei (1989) and Olver (1993) describe variational symmetries in considerable detail. These symmetries are most important in the context of PDEs, where (by Nöther's theorem) they can be used to derive conservation laws.

Some invariant solutions of ODEs have special topological properties, which distinguish them from neighbouring solutions. Section 3.6 of Bluman and Kumei (1989) includes many examples of such invariant solutions.

Exercises

4.1 Calculate the Lie point symmetry generators for the ODE

$$y'' = \frac{y'}{y^2}.$$

Use canonical coordinates corresponding to $X_1 = \partial_x$ to reduce this ODE to a simple first-order equation, and thereby solve the original ODE. Now try to solve the ODE by using another symmetry generator. What happens?

4.2 The ODE

$$y'' = \frac{3y'^2}{2y} + 2y^3$$

has (amongst others) the Lie symmetries generated by $X = \partial_x$. Use these symmetries to reduce the ODE to a first-order ODE that is solvable by a standard technique. Hence solve the original ODE.

4.3 The ODE $y'' = 0$ is the Euler–Lagrange equation for the variational problem whose Lagrangian is $L = \frac{1}{2}y'^2$. Use (3.28) to find three linearly independent generators of variational symmetries for this problem.

4.4 Derive the Euler–Lagrange equation for the variational problem with Lagrangian

$$L = \tfrac{1}{2}y'^2 - \frac{y^2}{8x^2} + \frac{1}{\sqrt{x}\,y} + \frac{1}{2y^2}.$$

Show that the scaling symmetries of the Euler–Lagrange equation are variational symmetries, and hence find the general solution of the variational problem.

4.5 Show that $X = xy\partial_x + y^2\partial_y$ generates Lie point symmetries of

$$y' = \frac{y^3}{(x+1)y^2 - x^2}.$$

Which solutions are invariant under these symmetries?

4.6 Find all solutions of (1.16) that are invariant under the rotational symmetries generated by $X = -y\partial_x + x\partial_y$. Use Fig. 1.5 to explain the topological significance of the result.

4.7 Find a nontrivial group-invariant solution to the *Thomas–Fermi equation*

$$y'' = x^{-\frac{1}{2}} y^{\frac{3}{2}}.$$

4.8 The *Poisson–Boltzman equation*,

$$y'' = \frac{-k}{x} y' - \delta e^y, \qquad k \neq 0, \quad \delta \in \{-1, 1\},$$

has Lie point symmetries generated by

$$X = x\partial_x - 2\partial_y.$$

This ODE arises from the variational problem with the Lagrangian

$$L = \frac{x^k y'^2}{2} - \delta x^k e^y.$$

Find the value of k for which X generates variational symmetries and solve the ODE in this case. For all other nonzero k, find all solutions that are invariant under the group generated by X.

4.9 Consider the family of variational problems with a Lagrangian of the form

$$L = \frac{x^2 y'^2}{2} - \frac{x^2 y^k}{k}, \qquad k \neq 0.$$

Find each value of k for which variational symmetries exist and hence solve the Euler–Lagrange equation in each case.

5

Lie Symmetries with Several Parameters

Because I could not bear to make
An Algebraist cry
I gazed with interest at X
And never thought of Why.

(G. K. Chesterton: True Sympathy)

5.1 Differential Invariants and Reduction of Order

A single generator of Lie point symmetries enables us to reduce the order of an ODE once. We have also seen an example of a double reduction of order using two Lie point symmetries. In fact, one can reduce an nth order ODE with $R \leq n$ Lie point symmetries to an ODE of order $n - R$ (or to an algebraic equation, if $R = n$). This section describes how such reductions are achieved.

If X generates Lie point symmetries of the ODE

$$y^{(n)} = \omega\big(x, y, y', \ldots, y^{(n-1)}\big), \qquad n \geq 2 \tag{5.1}$$

then, in terms of canonical coordinates (r, s), the ODE (5.1) reduces to

$$v^{(n-1)} = \Omega\big(r, v, \ldots, v^{(n-2)}\big), \tag{5.2}$$

where $v = v(r, \dot{s})$ is any function such that $v_{\dot{s}} \neq 0$. The reduced ODE (5.2) consists entirely of functions that are invariant under the (prolonged) action of the group generated by $X = \partial_s$. Such functions are called *differential invariants*. A nonconstant function $I(x, y, y', \ldots, y^{(k)})$ is a kth order differential invariant of the group generated by X if

$$X^{(k)} I = 0. \tag{5.3}$$

In canonical coordinates, $X^{(k)} = \partial_s$, so every kth order differential invariant is of the form

$$I = F\left(r, \dot{s}, \ldots, s^{(k)}\right)$$

or (equivalently)

$$I = F\left(r, v, \ldots, v^{(k-1)}\right) \tag{5.4}$$

for some function F. The invariant canonical coordinate $r(x, y)$ is the only differential invariant of order zero (up to functional dependence). All first-order differential invariants are functions of $r(x, y)$ and $v(x, y, y')$. Furthermore, all differential invariants of order 2 or greater are functions of r, v and derivatives of v with respect to r. Therefore, r and v are called *fundamental differential invariants*. We can usually find a convenient pair of fundamental differential invariants without first having to determine s. From (5.3), every kth order differential invariant satisfies

$$\xi I_x + \eta I_y + \cdots + \eta^{(k)} I_{y^{(k)}} = 0,$$

so (by the method of characteristics), I is a first integral of

$$\frac{dx}{\xi} = \frac{dy}{\eta} = \cdots = \frac{dy^{(k)}}{\eta^{(k)}}. \tag{5.5}$$

In particular, r is a first integral of

$$\frac{dx}{\xi} = \frac{dy}{\eta},$$

and v is a first integral of

$$\frac{dx}{\xi} = \frac{dy}{\eta} = \frac{dy'}{\eta^{(1)}}.$$

Sometimes it is necessary to use r to obtain v, as the following example shows.

Example 5.1 Find fundamental differential invariants of the group of rotations generated by

$$X = -y\partial_x + x\partial_y. \tag{5.6}$$

These satisfy $Xr = 0$ and $X^{(1)}v = 0$, so $r(x, y)$ is a first integral of

$$\frac{dx}{-y} = \frac{dy}{x};$$

one solution is $r = (x^2 + y^2)^{\frac{1}{2}}$. Similarly, $v(x, y, y')$ is a first integral of

$$\frac{dx}{-y} = \frac{dy}{x} = \frac{dy'}{1 + y'^2}. \tag{5.7}$$

For simplicity, let us restrict attention to the region $x > 0$, where

$$\frac{dy}{x} = \frac{dy}{(r^2 - y^2)^{\frac{1}{2}}}.$$

Then the first integrals of (5.7) are of the form

$$I = F\left(r,\ \tan^{-1} y' - \sin^{-1} \frac{y}{r}\right) = F\left(r,\ \tan^{-1} y' - \tan^{-1} \frac{y}{x}\right).$$

A convenient choice for v is

$$v = \tan\left(\tan^{-1} y' - \tan^{-1} \frac{y}{x}\right) = \frac{xy' - y}{x + yy'}.$$

It is left to the reader to verify that $v = r\dot{s}$, where $s = \tan^{-1} \frac{y}{x}$.

An ODE that has more than one Lie point symmetry generator can be written in terms of the differential invariants of each generator. Consequently, the ODE can be written in terms of functions that are invariant under all of its symmetry generators. Suppose that $\{X_1, \ldots, X_R\}$ is a basis of $\mathcal{L}$. The fundamental differential invariants of the group generated by $\mathcal{L}$ are solutions of the system

$$\begin{bmatrix} \xi_1 & \eta_1 & \eta_1^{(1)} & \ldots & \eta_1^{(R)} \\ \xi_2 & \eta_2 & \eta_2^{(1)} & \ldots & \eta_2^{(R)} \\ \vdots & \vdots & \vdots & & \vdots \\ \xi_R & \eta_R & \eta_R^{(1)} & \ldots & \eta_R^{(R)} \end{bmatrix} \begin{bmatrix} I_x \\ I_y \\ I_{y'} \\ \vdots \\ I_{y^{(R)}} \end{bmatrix} = \begin{bmatrix} 0 \\ 0 \\ \vdots \\ 0 \end{bmatrix}. \tag{5.8}$$

This system has two functionally independent solutions (provided that the matrix on the left-hand side has rank R). They can be found by using Gaussian elimination and the method of characteristics. One solution is independent of $y^{(R)}$ and is denoted r_R. We use v_R to denote the other solution, which depends nontrivially on $y^{(R)}$. As dv_R/dr_R depends on $y^{(R+1)}$, and so on, the ODE (5.1) reduces to

$$v_R^{(n-R)} = \Omega\left(r_R, v_R, \ldots, v_R^{(n-R-1)}\right), \qquad v_R^{(k)} \equiv \frac{d^k v_R}{dr_R^k},$$

for some function Ω. Thus an R-parameter symmetry group enables us to reduce the order of the ODE by R.

Example 5.2 The ODE

$$y^{(iv)} = \frac{2}{y}(1 - y')y''' \tag{5.9}$$

has a three-parameter Lie group of point symmetries, generated by

$$X_1 = \partial_x, \qquad X_2 = x\partial_x + y\partial_y, \qquad X_3 = x^2\partial_x + 2xy\partial_y. \tag{5.10}$$

The fundamental differential invariants are obtained by solving

$$\begin{bmatrix} 1 & 0 & 0 & 0 & 0 \\ x & y & 0 & -y'' & -2y''' \\ x^2 & 2xy & 2y & 2(y' - xy'') & -4xy''' \end{bmatrix} \begin{bmatrix} I_x \\ I_y \\ I_{y'} \\ I_{y''} \\ I_{y'''} \end{bmatrix} = \begin{bmatrix} 0 \\ 0 \\ 0 \end{bmatrix}.$$

First, simplify the problem by using Gaussian elimination:

$$\begin{bmatrix} 1 & 0 & 0 & 0 & 0 \\ 0 & y & 0 & -y'' & -2y''' \\ 0 & 0 & y & y' & 0 \end{bmatrix} \begin{bmatrix} I_x \\ I_y \\ I_{y'} \\ I_{y''} \\ I_{y'''} \end{bmatrix} = \begin{bmatrix} 0 \\ 0 \\ 0 \end{bmatrix}. \tag{5.11}$$

Then use each equation of (5.11) in turn to determine the differential invariants, as follows. The third equation, $yI_{y'} + y'I_{y''} = 0$, gives

$$I = I(x, y, 2yy'' - y'^2, y''').$$

Substituting this result into the second equation yields

$$I = I(x, 2yy'' - y'^2, y^2y''').$$

Finally, the first equation of (5.11) gives

$$I = I(2yy'' - y'^2,\ y^2y''').$$

So the fundamental differential invariants of the group generated by (5.10) are

$$r_3 = 2yy'' - y'^2, \qquad v_3 = y^2 y'''. \tag{5.12}$$

Having found the fundamental differential invariants, we can now calculate higher-order differential invariants, for example,

$$\frac{dv_3}{dr_3} = \frac{D_x v_3}{D_x r_3} = \frac{yy^{(iv)}}{2y'''} + y'.$$

Therefore the ODE (5.9) is equivalent to

$$\frac{dv_3}{dr_3} = 1,$$

whose general solution is

$$v_3 = r_3 + c_1.$$

This algebraic equation is equivalent to the third-order ODE

$$y''' = \frac{2yy'' - y'^2 + c_1}{y^2},$$

which is invariant under the three-parameter Lie group generated by $\mathcal{L}$.

Fundamental differential invariants can be used to construct ODEs that have given Lie point symmetries. If (r_R, v_R) are fundamental differential invariants of an R-dimensional Lie group G, then every ODE (5.1) of order $n \geq R$ that has G as its symmetry group can be written in the form

$$v_R^{(n-R)} = F\left(r_R, v_R, \ldots, v_R^{(n-1-R)}\right), \tag{5.13}$$

for some function F. By writing (5.13) in terms of $x, y, \ldots, y^{(n)}$, one obtains a family of ODEs that have the desired symmetries. (N.B. Some of these ODEs may have extra symmetries.)

Example 5.3 The fundamental differential invariants of the three-parameter group generated by

$$X_1 = \partial_x, \qquad X_2 = \partial_y, \qquad X_3 = x\partial_x + y\partial_y \tag{5.14}$$

are

$$r_3 = y', \qquad v_3 = \frac{y'''}{y''^2}.$$

Hence the most general third-order ODE with these symmetries is $v_3 = F(r_3)$, which amounts to

$$y''' = y''^2 F(y').$$

The most general fourth-order ODE with these symmetries is

$$y^{(iv)} = \frac{2y'''^2}{y''} + y''^3 F\left(y', \frac{y'''}{y''^2}\right),$$

which is equivalent to $\frac{dv_3}{dr_3} = F(r_3, v_3)$.

N.B. This method works for ODEs of order $n \geq R$. There may be ODEs of order $n < R$ whose symmetries include those in G. For instance, the symmetries of $y'' = 0$ include those generated by (5.14).

5.2 The Lie Algebra of Point Symmetry Generators

Suppose that the Lie point symmetries of the ODE (5.1) are generated by $\mathcal{L}$, which is R dimensional. By rewriting (5.1) in terms of the fundamental differential invariants (r_R, v_R), we obtain an ODE of order $n - R$. If the reduced ODE can be solved (as in Example 5.2), we are left with an algebraic equation,

$$v_R = F(r_R; c_1, \dots, c_{n-R}),$$

which is equivalent to an ODE of order R that has the R-parameter group of symmetries generated by $\mathcal{L}$. Is there a way to use these symmetries to complete the solution of the ODE? To answer this question, it is necessary to learn more about the structure of $\mathcal{L}$.

Suppose that $X_1, X_2 \in \mathcal{L}$, where

$$X_i = \xi_i(x, y)\partial_x + \eta_i(x, y)\partial_y, \qquad i = 1, 2. \tag{5.15}$$

The product $X_1 X_2$ is a second-order partial differential operator:

$$X_1 X_2 = \xi_1\xi_2\partial_x^2 + (\xi_1\eta_2 + \eta_1\xi_2)\partial_x\partial_y + \eta_1\eta_2\partial_y^2 + (X_1\xi_2)\partial_x + (X_1\eta_2)\partial_y.$$

The product $X_2 X_1$ is also second order, with exactly the same second-order terms as $X_1 X_2$. Therefore the *commutator* of X_1 with X_2,

$$[X_1, X_2] = X_1 X_2 - X_2 X_1, \tag{5.16}$$

is a first-order operator. Specifically,

$$[X_1, X_2] = (X_1\xi_2 - X_2\xi_1)\partial_x + (X_1\eta_2 - X_2\eta_1)\partial_y. \tag{5.17}$$

The commutator has many useful properties, some of which are obvious from the definition. It is *antisymmetric*, that is,

$$[X_2, X_1] = -[X_1, X_2], \tag{5.18}$$

and satisfies the *Jacobi identity*

$$[X_1, [X_2, X_3]] + [X_2, [X_3, X_1]] + [X_3, [X_1, X_2]] = 0. \tag{5.19}$$

The commutator is also *bilinear* (i.e., linear in both arguments):

$$\begin{aligned} [c_1X_1 + c_2X_2,\ X_3] &= c_1[X_1, X_3] + c_2[X_2, X_3], \\ [X_1,\ c_2X_2 + c_3X_3] &= c_2[X_1, X_2] + c_3[X_1, X_3]. \end{aligned} \tag{5.20}$$

(Here, as usual, c_i denotes an arbitrary constant.)

Under a change of variables from (x, y) to (u, v), each generator X_i transforms according to the chain rule. To find out how this transformation affects the commutator, let

$$X_3 = [X_1, X_2]. \tag{5.21}$$

For the time being, we use $\check{X}_i$ to denote the result of writing X_i in terms of (u, v):

$$\check{X}_i = (X_i u)\partial_u + (X_i v)\partial_v.$$

Let $F(u, v)$ be an arbitrary function. Then

$$\begin{aligned} [\check{X}_1, \check{X}_2]F &= \check{X}_1\big\{(X_2u)F_u + (X_2v)F_v\big\} - \check{X}_2\big\{(X_1u)F_u + (X_1v)F_v\big\} \\ &= (X_1X_2u)F_u + (X_1X_2v)F_v - (X_2X_1u)F_u - (X_2X_1v)F_v \\ &= \big([X_1, X_2]u\big)F_u + \big([X_1, X_2]v\big)F_v \\ &= (X_3u)F_u + (X_3v)F_v. \end{aligned}$$

However F is arbitrary, so

$$[\check{X}_1, \check{X}_2] = (X_3u)\partial_u + (X_3v)\partial_v = \check{X}_3. \tag{5.22}$$

Therefore the commutator of X_1 with X_2 is essentially independent of coordinate system in which it is calculated. There is no longer any need to distinguish between $\check{X}_i$ and X_i, so we revert to using X_i to denote the generators (in every coordinate system).

So far, we have considered the commutator of generators acting on the plane. The commutator of the prolonged generators

$$X_i^{(k)} = \xi\partial_x + \eta\partial_y + \eta^{(1)}\partial_{y'} + \cdots + \eta^{(k)}\partial_{y^{(k)}}$$

is defined similarly:

$$\left[X_1^{(k)}, X_2^{(k)}\right] = X_1^{(k)}X_2^{(k)} - X_2^{(k)}X_1^{(k)}.$$

We now show that if $[X_1, X_2] = X_3$ then

$$\left[X_1^{(k)}, X_2^{(k)}\right] = X_3^{(k)}. \tag{5.23}$$

Suppose that

$$X_1 = \partial_y, \qquad X_2 = \xi(x, y)\partial_x + \eta(x, y)\partial_y. \tag{5.24}$$

There is no loss of generality in making this supposition, for we have shown that a change of variables – in this case, to canonical coordinates of X_1 – does not affect the commutator. From (5.24),

$$X_3 = [X_1, X_2] = \xi_y\partial_x + \eta_y\partial_y.$$

Therefore the prolongation formula gives

$$X_3^{(1)} = \xi_y\partial_x + \eta_y\partial_y + (D_x\eta_y - y'D_x\xi_y)\partial_{y'}.$$

Because ∂_y and D_x commute, that is,

$$D_x\partial_y = \partial_x\partial_y + y'\partial_y^2 + \ldots = \partial_y D_x,$$

the last term of $X_3^{(1)}$ may be rewritten as $\eta_y^{(1)}\partial_{y'}$, where

$$\eta^{(1)} = D_x\eta - y'D_x\xi.$$

Therefore

$$\begin{aligned} X_3^{(1)} &= \xi_y\partial_x + \eta_y\partial_y + \eta_y^{(1)}\partial_{y'} \\ &= \left[\partial_y, \xi\partial_x + \eta\partial_y + \eta^{(1)}\partial_{y'}\right] \\ &= \left[X_1^{(1)}, X_2^{(1)}\right]. \end{aligned}$$

So the result (5.23) holds for $k = 1$. The result for $k > 1$ is obtained similarly, using the observation that

$$D_x \eta_y^{(k-1)} - y^{(k)} D_x \xi_y = \partial_y \left(D_x \eta^{(k-1)} - y^{(k)} D_x \xi \right) = \eta_y^{(k)}.$$

Furthermore, (5.23) holds in any system of coordinates, because it holds in one system.

We have not yet used the fact that X_1 and X_2 generate Lie point symmetries of the ODE (5.1). Indeed, everything that we have discussed applies to all first-order partial differential operators of the form (5.15). However each generator in $\mathcal{L}$ also satisfies the linearized symmetry condition

$$X_i^{(n)} \left(y^{(n)} - \omega \right) = \eta_i^{(n)} - X_i^{(n)} \omega = 0 \qquad \text{when} \quad y^{(n)} = \omega.$$

For $n \geq 2$, the prolongation formula implies that $\eta_i^{(n)}$ is linear in the highest derivative, $y^{(n)}$, whereas ω, and thus $X_i^{(n)} \omega$, is independent of the highest derivative. Therefore the linearized symmetry condition is satisfied if and only if

$$X_i^{(n)} \left(y^{(n)} - \omega \right) = \lambda_i \left(y^{(n)} - \omega \right), \tag{5.25}$$

where

$$\lambda_i \left(x, y, y', \ldots, y^{(n-1)} \right) = \frac{\partial \eta_i^{(n)}}{\partial y^{(n)}}.$$

This alternative characterization of the linearized symmetry condition leads to an important result: if X_1 and X_2 generate Lie point symmetries, then so does $X = [X_1, X_2]$. This result is obtained quite simply, as follows. For brevity, let $\Delta = y^{(n)} - \omega$. Then, from (5.23) and (5.25),

$$\begin{aligned} X^{(n)} \Delta &= \left[X_1^{(n)}, X_2^{(n)} \right] \Delta \\ &= X_1^{(n)} (\lambda_2 \Delta) - X_2^{(n)} (\lambda_1 \Delta) \\ &= \left(X_1^{(n)} \lambda_2 - X_2^{(n)} \lambda_1 \right) \Delta. \end{aligned}$$

Hence $X^{(n)} \Delta = 0$ when $\Delta = 0$, and therefore X generates Lie point symmetries. [The same result also holds for $n = 1$, but the above argument requires slight modification because $\eta_i^{(1)}$ is quadratic in y'.]

For $n \geq 2$, the set $\mathcal{L}$ is a finite-dimensional vector space. Once we have chosen a basis $\{X_1, \ldots, X_R\}$ for $\mathcal{L}$, every generator of Lie point symmetries can be written as a linear combination of the generators in the basis. The commutator is bilinear, and so it is sufficient to restrict attention to the commutators of the basis generators. We have just demonstrated that $\mathcal{L}$ is closed under the commutator, that is,

$$X_i, X_j \in \mathcal{L} \Rightarrow [X_i, X_j] \in \mathcal{L}.$$

Therefore the commutator of any two generators in the basis is a linear combination of the basis generators:

$$[X_i, X_j] = c_{ij}^k X_k. \tag{5.26}$$

(Remember, we sum over all possible values of any index that occurs twice.) The constants c_{ij}^k are called *structure constants*. If $[X_i, X_j] = 0$, the generators X_i and X_j are said to *commute*. In particular, every generator commutes with itself.

Example 5.4 The vector space of generators of Lie point symmetries of $y''' = y^{-3}$ is two dimensional and is spanned by

$$X_1 = \partial_x, \qquad X_2 = x\partial_x + \tfrac{3}{4}y\partial_y. \tag{5.27}$$

Therefore the commutator of X_1 with X_2 is

$$\begin{aligned}[X_1, X_2] &= \big(X_1(x) - X_2(1)\big)\partial_x + \Big(X_1\big(\tfrac{3}{4}y\big) - X_2(0)\Big)\partial_y \\ &= \partial_x \\ &= X_1.\end{aligned}$$

The remaining commutators are found without any need for further calculation. Each generator commutes with itself, so

$$[X_1, X_1] = 0, \qquad [X_2, X_2] = 0.$$

Moreover, the commutator is antisymmetric, and hence

$$[X_2, X_1] = -[X_1, X_2] = -X_1.$$

Summarizing these results, the only nonzero structure constants for the basis (5.27) are

$$c_{12}^1 = 1, \qquad c_{21}^1 = -1. \tag{5.28}$$

The existence of the commutator as a "product" on $\mathcal{L}$ means that $\mathcal{L}$ is not just a vector space; it is a *Lie algebra*. Formally, a Lie algebra is a vector space that is closed under a product $[\cdot, \cdot]$ which is bilinear, antisymmetric, and satisfies the Jacobi identity. The last two conditions impose some constraints on the structure constants. The commutator is antisymmetric if

$$[X_j, X_i] = -[X_i, X_j], \qquad \forall\, i, j, \tag{5.29}$$

which implies that

$$c_{ij}^k = -c_{ji}^k, \qquad \forall\, i, j, k. \tag{5.30}$$

One consequence of this identity is that we need only calculate the commutators of basis generators with $i < j$. The Jacobi identity

$$[X_i, [X_j, X_k]] + [X_j, [X_k, X_i]] + [X_k, [X_i, X_j]] = 0, \qquad \forall\, i, j, k, \tag{5.31}$$

holds if and only if

$$c_{ij}^q c_{kq}^l + c_{jk}^q c_{iq}^l + c_{ki}^q c_{jq}^l = 0, \qquad \forall\, i, j, k, l. \tag{5.32}$$

Lie algebras occur in many branches of applied mathematics and physics. Commonly, one is interested in the action of a multiparameter Lie group, whose linearization about the identity yields the Lie algebra that generates the group. Lie algebras may also exist without reference to an underlying Lie group. A Lie algebra is defined abstractly by its structure constants, but it may appear in many different forms (or realizations), as the following example shows.

Example 5.5 Perhaps the most well-known Lie algebra is the space of vectors $\mathbf{x} \in \mathbb{R}^3$ under the cross product, which plays the rôle of the commutator:

$$[\mathbf{x}_1, \mathbf{x}_2] = \mathbf{x}_1 \times \mathbf{x}_2.$$

(The cross product is bilinear and antisymmetric; the reader may wish to check that the Jacobi identity is satisfied.) With the standard Cartesian basis for $\mathbb{R}^3$,

$$\mathbf{x}_1 = (1\ 0\ 0)^{\mathrm{T}}, \qquad \mathbf{x}_2 = (0\ 1\ 0)^{\mathrm{T}}, \qquad \mathbf{x}_3 = (0\ 0\ 1)^{\mathrm{T}},$$

the cross product gives the following nontrivial relations:

$$\mathbf{x}_1 \times \mathbf{x}_2 = \mathbf{x}_3, \qquad \mathbf{x}_1 \times \mathbf{x}_3 = -\mathbf{x}_2, \qquad \mathbf{x}_2 \times \mathbf{x}_3 = \mathbf{x}_1.$$

The only nonzero structure constants are

$$c_{12}^3 = c_{23}^1 = c_{31}^2 = 1, \qquad c_{21}^3 = c_{32}^1 = c_{13}^2 = -1; \tag{5.33}$$

note that the structure constants are unchanged by cyclic permutations of the indices (123). The abstract Lie algebra which has the structure constants (5.33) in some basis is called $\mathfrak{so}(3)$. The Lie group generated by $\mathfrak{so}(3)$ is the *special orthogonal group SO*(3), a simple example of which is the group of rotations in $\mathbb{R}^3$. The Lie algebra $\mathfrak{so}(3)$ can also be realized in terms of generators of Lie point transformations of the plane:

$$X_1 = y\partial_x - x\partial_y, \qquad X_2 = \tfrac{1}{2}(1 + x^2 - y^2)\partial_x + xy\partial_y,$$
$$X_3 = xy\partial_x + \tfrac{1}{2}(1 - x^2 + y^2)\partial_y. \tag{5.34}$$

The reader should check that this basis gives the structure constants (5.33).

From our earlier results, the structure constants are unaffected by either prolongation or a change of variables. However, they do depend on the choice of basis for $\mathcal{L}$, and it is useful to try to choose a basis with as few nonzero structure constants as possible. If all generators in the basis commute (i.e., every structure constant is zero), the Lie algebra is *abelian*.

Example 5.6 Consider the most general two-dimensional Lie algebra, with a basis $\{X_1, X_2\}$. The commutator of X_1 with X_2 is of the form

$$[X_1, X_2] = c_{12}^1 X_1 + c_{12}^2 X_2. \tag{5.35}$$

The Lie algebra is abelian if and only if $c_{12}^1 = c_{12}^2 = 0$. Otherwise, there exists a basis $\{\tilde{X}_1, \tilde{X}_2\}$ such that

$$[\tilde{X}_1, \tilde{X}_2] = \tilde{X}_1. \tag{5.36}$$

To find this basis, note that the commutator of any two generators is necessarily a multiple of the right-hand side of (5.35). So let

$$\tilde{X}_1 = c_{12}^1 X_1 + c_{12}^2 X_2.$$

If $c_{12}^1 \neq 0$, then X_2 is linearly independent of $\tilde{X}_1$, and

$$[\tilde{X}_1, X_2] = c_{12}^1 [X_1, X_2] = c_{12}^1 \tilde{X}_1.$$

Rescaling, we obtain (5.36) by choosing

$$\tilde{X}_2 = \frac{1}{c_{12}^1} X_2.$$

Similarly, if $c^1_{12} = 0$ but the Lie algebra is non-abelian, we may satisfy (5.36) by taking

$$\tilde{X}_2 = \frac{-1}{c^2_{12}} X_1.$$

So every two-dimensional Lie algebra is either abelian or else can be written in a basis for which the only nonzero structure constants are

$$c^1_{12} = 1, \qquad c^1_{21} = -1. \tag{5.37}$$

Ordinary vector spaces are constructed from subpaces. Similarly, Lie algebras are built from subalgebras; the way in which these are joined together determines the structure constants. For convenience, let $[\mathcal{M}, \mathcal{N}]$ denote the set of all commutators of generators in $\mathcal{M} \subset \mathcal{L}$ with generators in $\mathcal{N} \subset \mathcal{L}$, that is,

$$[\mathcal{M}, \mathcal{N}] = \{[X_i, X_j] : X_i \in \mathcal{M},\ X_j \in \mathcal{N}\}. \tag{5.38}$$

A subspace $\mathcal{M} \subset \mathcal{L}$ is a *subalgebra* if it is closed under the commutator:

$$[\mathcal{M}, \mathcal{M}] \subset \mathcal{M}. \tag{5.39}$$

In other words, any subalgebra of a Lie algebra is a Lie algebra in its own right. A subalgebra $\mathcal{M} \subset \mathcal{L}$ is an *ideal* of $\mathcal{L}$ if

$$[\mathcal{M}, \mathcal{L}] \subset \mathcal{M}. \tag{5.40}$$

Trivially, $\{0\}$ and $\mathcal{L}$ are subalgebras; moreover, they are ideals of $\mathcal{L}$. Any ideal other than $\{0\}$ and $\mathcal{L}$ is called a *proper ideal*. If a Lie algebra is non-abelian and has no proper ideals, it is said to be *simple*. Every one-dimensional subspace of $\mathcal{L}$ is a subalgebra (but not necessarily an ideal), because each generator commutes with itself. Almost all Lie algebras of dimension $R \geq 2$ have at least one two-dimensional subalgebra; the one exception is the simple Lie algebra $\mathfrak{so}(3)$.

Example 5.7 Consider the three-dimensional Lie algebra with the basis

$$X_1 = \partial_x, \qquad X_2 = x\partial_x, \qquad X_3 = x^2\partial_x. \tag{5.41}$$

The nontrivial commutators $[X_i, X_j]$ (with $i < j$) are

$$[X_1, X_2] = X_1, \qquad [X_1, X_3] = 2X_2, \qquad [X_2, X_3] = X_3. \tag{5.42}$$

The abstract Lie algebra with the commutators (5.42) is called $\mathfrak{sl}(2)$; it generates the *special linear group* $SL(2)$. The subspace spanned by X_1 and X_2 is a two-dimensional subalgebra, as is Span(X_2, X_3). However Span(X_1, X_3) is not a subalgebra, because $[X_1, X_3]$ does not lie in this subspace. Although $\mathfrak{sl}(2)$ has nontrivial subalgebras, it has no ideals other than itself and $\{0\}$; therefore it is simple.

Given any Lie algebra $\mathcal{L}$, one ideal that can always be constructed is the *derived subalgebra* $\mathcal{L}^{(1)}$, which consists of all commutators of elements of $\mathcal{L}$:

$$\mathcal{L}^{(1)} = [\mathcal{L}, \mathcal{L}]. \tag{5.43}$$

Clearly, $[\mathcal{L}^{(1)}, \mathcal{L}]$ is a subset of $\mathcal{L}^{(1)}$, which is why $\mathcal{L}^{(1)}$ is an ideal.

If $\mathcal{L}^{(1)} \neq \mathcal{L}$, we can go on to find the derived subalgebra of $\mathcal{L}^{(1)}$, namely

$$\mathcal{L}^{(2)} = \left[\mathcal{L}^{(1)}, \mathcal{L}^{(1)}\right]. \tag{5.44}$$

We can continue this process, letting

$$\mathcal{L}^{(k)} = \left[\mathcal{L}^{(k-1)}, \mathcal{L}^{(k-1)}\right], \tag{5.45}$$

until we fail to obtain a new subalgebra. If the series of derived subalgebras terminates with $\mathcal{L}^{(k)} = \{0\}$ for some k, then $\mathcal{L}$ is said to be *solvable*. Equivalently, an R-dimensional Lie algebra is solvable if there is a chain of subalgebras

$$\{0\} = \mathcal{L}_0 \subset \mathcal{L}_1 \subset \cdots \subset \mathcal{L}_R = \mathcal{L}, \tag{5.46}$$

where $\dim(\mathcal{L}_k) = k$, such that $\mathcal{L}_{k-1}$ is an ideal of $\mathcal{L}_k$ for each k. Every abelian Lie algebra is solvable, whereas simple Lie algebras are not solvable. We have seen that any non-abelian two-dimensional Lie algebra has a basis such that $[X_1, X_2] = X_1$, and hence is solvable. For each $R \geq 3$, there exist Lie algebras that are not solvable.

Given an R-dimensional solvable Lie algebra, it is convenient to choose a basis such that

$$X_k \in \mathcal{L}_k, \qquad X_k \notin \mathcal{L}_{k-1}, \qquad k = 1, \ldots, R, \tag{5.47}$$

and hence

$$\mathcal{L}_k = \text{Span}(X_1, \ldots, X_k).$$

We shall call any basis defined by (5.47) a *canonical basis*. Equivalently, a basis is canonical if the structure constants satisfy

$$c_{ij}^{k} = 0, \qquad \forall\, i < j \leq k. \tag{5.48}$$

Example 5.8 The fourth-order ODE

$$y^{(iv)} = y'''^{\frac{4}{3}} \tag{5.49}$$

has a five-dimensional Lie algebra $\mathcal{L}$, which has a basis

$$\begin{gathered} X_1 = \partial_y, \qquad X_2 = x\partial_y, \qquad X_3 = x^2\partial_y, \\ X_4 = \partial_x, \qquad X_5 = x\partial_x. \end{gathered} \tag{5.50}$$

The nonzero commutators $[X_i, X_j]$ (with $i < j$) are

$$\begin{gathered} [X_2, X_4] = -X_1, \qquad [X_2, X_5] = -X_2, \qquad [X_3, X_4] = -2X_2, \\ [X_3, X_5] = -2X_3, \qquad [X_4, X_5] = X_4. \end{gathered} \tag{5.51}$$

Therefore $\mathcal{L}$ has the derived algebra

$$\mathcal{L}^{(1)} = \mathrm{Span}(X_1, X_2, X_3, X_4). \tag{5.52}$$

As $\mathcal{L}^{(1)} \neq \mathcal{L}$, we may go on to calculate $\mathcal{L}^{(2)}$. From (5.51), the nonzero commutators of elements in $\mathcal{L}^{(1)}$ are

$$[X_2, X_4] = -X_1, \qquad [X_3, X_4] = -2X_2.$$

Therefore

$$\mathcal{L}^{(2)} = \mathrm{Span}(X_1, X_2). \tag{5.53}$$

Finally, X_1 and X_2 commute, so the series terminates with

$$\mathcal{L}^{(3)} = \{0\}. \tag{5.54}$$

Hence $\mathcal{L}$ is solvable. Note that the basis (5.50) is canonical.

5.3 Stepwise Integration of ODEs

We now return to the problem of integrating an ODE of order R that has an R-dimensional Lie algebra $\mathcal{L}$. The ODE can be written in terms of the fundamental differential invariants as

$$v_R = F(r_R), \tag{5.55}$$

for some function F. We aim to solve the ODE by using each symmetry generator in turn. (N.B. Henceforth, we do not explicitly refer to the order of prolongation of the generators, unless there is a good reason for doing so. Instead, we adopt the convention that generators are prolonged sufficiently to describe the linearized group action on all variables.)

Suppose that the generators $X_1, \ldots, X_{R-1}$ form a subalgebra of L. Let (r_{R-1}, v_{R-1}) be the fundamental differential invariants of this subalgebra. If the remaining generator, X_R, acts on (r_{R-1}, v_{R-1}) as a generator of point transformations, there exist canonical coordinates

$$(r_R, s_R) = \big(r_R(r_{R-1}, v_{R-1}),\ s_R(r_{R-1}, v_{R-1})\big),$$

at every noninvariant "point," in terms of which $X_R = \partial_{s_R}$. (We already know r_R, and s_R can be found by the usual method.) Then v_R is a function of r_R and $\dot{s}_R \equiv ds_R/dr_R$ only, and so (5.55) can be inverted (at least, in principle) to yield

$$\dot{s}_R = G(r_R),$$

for some function G. Hence we obtain

$$s_R(r_{R-1}, v_{R-1}) = \int^{r_R(r_{R-1}, v_{R-1})} G(r_R)\, dr_R + c,$$

which is invariant under the group generated by $X_1, \ldots, X_{R-1}$. If this equation can be solved for v_{R-1} as a function of r_{R-1}, we obtain a problem of the form (5.55), but with $R-1$ replacing R. Provided that we can iterate this method sufficiently many times, we will obtain the general solution of the ODE.

N.B. Throughout this chapter, we focus on the problem of finding the general solution of an ODE. Solutions that are invariant under a one-parameter group can be sought with the methods described at the end of Chapter 4. Although there may be many invariant solutions, the Lie algebra can be used to classify them, as described in Chapter 10. This greatly reduces the effort needed to find all invariant solutions.

Clearly, X_R acts on (r_{R-1}, v_{R-1}) as a generator of point transformations if the restriction of X_R to the variables (r_{R-1}, v_{R-1}) is of the form

$$X_R = \alpha(r_{R-1}, v_{R-1})\partial_{r_{R-1}} + \beta(r_{R-1}, v_{R-1})\partial_{v_{R-1}}$$

for some functions α, β, at least one of which is nonzero. Therefore we require that

$$X_R\, r_{R-1} = \alpha(r_{R-1}, v_{R-1}), \qquad X_R\, v_{R-1} = \beta(r_{R-1}, v_{R-1}).$$

The differential invariants r_{R-1}, v_{R-1} satisfy

$$X_i\, r_{R-1} = 0, \qquad X_i\, v_{R-1} = 0, \qquad \forall\, i = 1, \ldots, R-1,$$

and hence

$$[X_i, X_R]\, r_{R-1} = X_i \alpha(r_{R-1}, v_{R-1}) = 0, \qquad \forall\, i = 1, \ldots, R-1.$$

This can be rewritten as

$$c^k_{iR} X_k r_{R-1} = 0, \qquad \forall\, i = 1, \ldots, R-1,$$

which leads to the condition

$$c^R_{iR}\alpha(r_{R-1}, v_{R-1}) = 0, \qquad \forall\, i = 1, \ldots, R-1.$$

By a similar argument, $[X_i, X_R]\, v_{R-1} = 0$ leads to the condition

$$c^R_{iR}\beta(r_{R-1}, v_{R-1}) = 0, \qquad \forall\, i = 1, \ldots, R-1.$$

Hence, since at least one of α, β is nonzero,

$$c^R_{iR} = 0, \qquad \forall\, i = 1, \ldots, R-1.$$

$\mathrm{Span}(X_1, \ldots, X_{R-1})$ is a subalgebra if

$$c^R_{ij} = 0, \qquad \forall\, 1 \le i < j \le R-1.$$

So X_R acts as a generator of point transformations on (r_{R-1}, v_{R-1}) if and only if

$$c^R_{ij} = 0, \qquad \forall\, 1 \le i < j \le R. \tag{5.56}$$

This condition enables us to reduce the order once. Similarly, a second reduction of order is possible if

$$c^{R-1}_{ij} = 0, \qquad \forall\, 1 \le i < j \le R-1. \tag{5.57}$$

Continuing in the same way, each generator X_k may be used to carry out one integration if

$$c^k_{ij} = 0, \qquad \forall\, 1 \le i < j \le k. \tag{5.58}$$

This condition is satisfied (in any canonical basis) if and only if $\mathcal{L}$ is a solvable Lie algebra.

So far, we have restricted attention to Lie algebras with $R \le n$. However, the above method also works if $R > n$, provided that $\mathcal{L}$ has an n-dimensional solvable subalgebra. In the next chapter, we shall put this method into practice and discuss what can be done if $\mathcal{L}$ does not have a sufficiently large solvable subalgebra.

Further Reading

Differential invariants are not only useful for reduction of order. They are also used to construct models that have given symmetries. Olver (1995) includes a clear, detailed description of differential invariants and some of their applications.

Lie algebras are widely used throughout mathematics and theoretical physics. Sattinger and Weaver (1986) is a straightforward introduction. For a more comprehensive treatment, I recommend Fuchs and Schweigert (1997).

Exercises

5.1 What is the most general third-order ODE whose symmetries include the group generated by $X = x\partial_x + \alpha y\partial_y$ (where α is a constant)?

5.2 Derive a set of fundamental differential invariants for
   (a) the group generated by $X_1 = xy\partial_x + y^2\partial_y$;
   (b) the group generated by $X_2 = x\partial_x - y\partial_y$;
   (c) the group generated by X_1 and X_2.

   Now find the most general third-order ODE whose symmetries include the group generated by X_1 and X_2.

5.3 Find the most general second-order ODE whose symmetries include the group generated by (5.34).

5.4 Show that the basis (5.34) has the $\mathfrak{so}(3)$ structure constants (5.33).

5.5 Show that there is no three-dimensional Lie algebra with the following commutators between its basis generators:

$$[X_1, X_2] = X_1, \qquad [X_1, X_3] = -X_3, \qquad [X_2, X_3] = X_2.$$

5.6 Consider the two-dimensional vector space spanned by the generators $X_1 = x\partial_x - \alpha y\partial_y$ and $X_2 = -y\partial_x + x\partial_y$. For which value of α is this vector space a Lie algebra? For every other $\alpha \in \mathbb{R}$, find the smallest Lie algebra $\mathcal{L}(\alpha)$ that contains X_1 and X_2. Show that $\mathcal{L}(\alpha)$ has an $\mathfrak{sl}(2)$ subalgebra, and state the dimension of the largest solvable subalgebras.

5.7 Show that if I is a differential invariant of the one-parameter groups generated by X_1 and X_2, it is also a differential invariant of the group generated by $[X_1, X_2]$. Hence find the lowest-order differential invariant that is common to the groups generated by $X_1 = \partial_y$ and $X_2 = 2xy\partial_x + y^2\partial_y$.

6

Solution of ODEs with Multiparameter Lie Groups

Little by little does the trick.

(Aesop: Fables)

6.1 The Basic Method: Exploiting Solvability

We now have a systematic method for solving ODEs with a sufficiently large solvable Lie (sub)algebra:

(1) Use the linearized symmetry condition to determine the Lie point symmetries.
(2) Calculate the commutators of the basis generators and hence find the series of derived subalgebras.
(3) Find a sufficiently large solvable subalgebra, choose a canonical basis, and calculate the fundamental differential invariants.
(4) Rewrite the ODE in terms of differential invariants; then use each generator in turn to carry out one integration, as described in Chapter 5.

The purpose of this section is to show how the method works in practice. In the following, we shall focus on step (4), having already obtained a canonical basis for the generators.

Example 6.1 Recall that the ODE

$$y'' = \frac{y'^2}{y} - y^2, \qquad y > 0, \tag{6.1}$$

has Lie point symmetries generated by

$$X_1 = \partial_x, \qquad X_2 = x\partial_x - 2y\partial_y. \tag{6.2}$$

The group generated by X_1 has fundamental differential invariants

$$r_1 = y, \qquad v_1 = y',$$

and the fundamental differential invariants of the group generated by X_1 and X_2 are

$$r_2 = \frac{y'}{y^{3/2}}, \qquad v_2 = \frac{y''}{y^2}.$$

Hence the ODE (6.1) reduces to the algebraic equation

$$v_2 = r_2^2 - 1. \tag{6.3}$$

The Lie algebra is solvable and (6.2) is a canonical basis, so X_2 generates point transformations of the variables (r_1, v_1). Explicitly,

$$X_2 r_1 = -2y = -2r_1, \qquad X_2^{(1)} v_1 = -3y' = -3v_1,$$

and so the restricition of X_2 to (r_1, v_1) is

$$X_2 = -2r_1 \partial_{r_1} - 3v_1 \partial_{v_1}.$$

We have already chosen the invariant canonical coordinate $r_2 = v_1 / r_1^{3/2}$; for simplicity, let $s_2 = -\frac{1}{2}\ln(r_1)$. Then

$$\frac{ds_2}{dr_2} = \frac{ds_2}{dx} \Big/ \frac{dr_2}{dx} = \frac{r_2}{3r_2^2 - 2v_2}.$$

Therefore, from (6.3), the reduced ODE is equivalent to

$$\frac{ds_2}{dr_2} = \frac{r_2}{r_2^2 + 2}.$$

The quadrature is straightforward:

$$s_2 = \tfrac{1}{2}\ln\left(r_2^2 + 2\right) + c.$$

After rewriting this solution in terms of (r_1, v_1), we obtain the algebraic equation

$$v_1 = \pm r_1 \left(4c_1^2 - 2r_1\right)^{\frac{1}{2}}, \tag{6.4}$$

where c_1 is an arbitrary positive constant. Having completed one step of the iteration, we repeat the same process, now using the generator X_1 to solve (6.4). With the canonical coordinates $(r_1, s_1) = (y, x)$, we obtain

$$\frac{ds_1}{dr_1} = \frac{1}{y'} = \frac{\pm 1}{r_1\left(4c_1^2 - 2r_1\right)^{1/2}}.$$

Therefore the general solution of the ODE is

$$s_1 = c_2 \mp c_1^{-1} \cosh^{-1}\left(c_1 \sqrt{\frac{2}{r_1}} \right).$$

Reverting to the original variables, we obtain

$$y = 2c_1^2 \operatorname{sech}^2\big(c_1(x - c_2)\big). \tag{6.5}$$

Example 6.2 Consider the third-order ODE

$$y''' = \frac{y''^2}{y'(1 + y')}, \tag{6.6}$$

whose Lie point symmetries are generated by

$$X_1 = \partial_x, \qquad X_2 = \partial_y, \qquad X_3 = x\partial_x + y\partial_y. \tag{6.7}$$

Fundamental differential invariants for each subalgebra $\mathcal{L}_k$ in the solvable chain are

$$\begin{aligned} &r_1 = y, && v_1 = y' && \text{for} \quad \mathcal{L}_1 = \mathrm{Span}(X_1); \\ &r_2 = y', && v_2 = y'' && \text{for} \quad \mathcal{L}_2 = \mathrm{Span}(X_1, X_2); \\ &r_3 = y', && v_3 = y'''/y''^2 && \text{for} \quad \mathcal{L}_3 = \mathrm{Span}(X_1, X_2, X_3). \end{aligned} \tag{6.8}$$

Thus the ODE (6.6) is equivalent to the algebraic equation

$$v_3 = \frac{1}{r_3(1 + r_3)}. \tag{6.9}$$

To find the restriction of X_3 to the differential invariants (r_2, v_2), note that

$$X_3^{(1)} r_2 = 0, \qquad X_3^{(2)} v_2 = -y'' = -v_2;$$

hence the restricted generator is

$$X_3 = -v_2 \partial_{v_2}.$$

Let

$$s_3 = -\ln|v_2| = -\ln|y''|;$$

then (6.9) is equivalent to the first-order ODE

$$\frac{ds_3}{dr_3} = -v_3 = \frac{-1}{r_3(1 + r_3)}.$$

Writing the solution of this ODE in terms of r_2 and v_2, we obtain

$$v_2 = \frac{c_1 r_2}{1 + r_2}. \tag{6.10}$$

In the next step, we find that the restriction of X_2 to the variables (r_1, v_1) is

$$X_2 = \partial_{r_1}.$$

Then $s_2 = r_1$ is a suitable canonical coordinate. Provided that $c_1 \neq 0$, the algebraic equation (6.10) is equivalent to

$$\frac{ds_2}{dr_2} = \frac{r_2}{v_2} = \frac{1}{c_1}(1 + r_2),$$

whose solution is (in terms of r_1, v_1)

$$r_1 = c_2 + \frac{1}{2c_1}(1 + v_1)^2. \tag{6.11}$$

Finally, with $s_1 = x$, we obtain

$$\frac{ds_1}{dr_1} = \frac{1}{v_1} = \frac{1}{-1 \pm \sqrt{2c_1(r_1 - c_2)}}.$$

After carrying out the quadrature and replacing s_1 and r_1 by x and y respectively, we obtain the general solution of (6.6) in closed form:

$$x = c_3 + \frac{1}{c_1}\left(\ln\left|-1 \pm \sqrt{2c_1(y - c_2)}\right| \pm \sqrt{2c_1(y - c_1)}\right). \tag{6.12}$$

Given a basis for an n-dimensional solvable subalgebra, it is common for there to be several different orderings of the generators, each of which is a canonical basis. In principle, it does not matter which canonical basis is chosen. In practice, some choices may make the quadratures unnecessarily difficult (or even intractable). It seems that the only way to resolve this problem is by experimenting with various canonical bases, as the next example shows. By now the procedure for solving ODEs should be familiar, so most details of the following calculations are omitted (they should be supplied by the reader).

Example 6.3 In the previous example, we could have chosen the canonical basis

$$X_1 = \partial_y, \qquad X_2 = \partial_x, \qquad X_3 = x\partial_x + y\partial_y, \tag{6.13}$$

which puts the commuting generators ∂_x and ∂_y in a different order from (6.7). The set of fundamental differential invariants (6.8) is almost unaltered; the only difference is that now $r_1 = x$. Hence the calculations are unchanged as far as (6.10), which now reduces to

$$\frac{ds_2}{dr_2} = \frac{1}{v_2} = \frac{1}{c_1}\left(1 + \frac{1}{r_2}\right).$$

The quadrature yields

$$r_1 = c_2 + \frac{1}{c_1}\left(\ln|v_1| + v_1\right), \tag{6.14}$$

which presents us with a problem: we cannot use (6.14) to obtain v_1 as a function of r_1. Therefore the ordering of the canonical basis turns out to be crucial in determining the general solution of (6.6).

The above examples are straightforward, because it is easy to find a canonical basis such that v_k can be written as a function of r_k at each stage. If such a basis cannot be found, it may be possible to obtain the general solution in parametric form. Suppose that

$$r = f(v) \tag{6.15}$$

and that $\dot{s} = ds/dr$ can be written in terms of r and v. Then

$$s = g(v) \equiv \int \dot{s}(r, v)\bigg|_{r=f(v)} \frac{df(v)}{dv}\, dv + c. \tag{6.16}$$

If (6.15) and (6.16) permit the parameter v to be written as a function of r and s, then v can be eliminated. This process, which is called *implicitization*, can always be carried out if $f(v)$ and $g(v)$ are rational polynomials. However not all ODEs allow implicitization; if v cannot be eliminated, the solution must remain in parametric form.

Example 6.4 Consider the ODE

$$y'' = \frac{y'^3}{y'^3 - 2}, \tag{6.17}$$

whose Lie algebra of Lie point symmetry generators is spanned by

$$X_1 = \partial_x, \qquad X_2 = \partial_y.$$

Using the fundamental differential invariants

$$(r_1, v_1) = (y, y'), \qquad (r_2, v_2) = (y', y'')$$

and the canonical coordinate $s_2 = r_1$, we obtain

$$r_1 = \frac{v_1^2}{2} + \frac{2}{v_1} + c_1. \tag{6.18}$$

Let $s_1 = x$, so that

$$\dot{s}_1 = \frac{ds_1}{dr_1} = \frac{1}{v_1}.$$

Then, from (6.16),

$$s_1 = v_1 + \frac{1}{v_1^2} + c_2. \tag{6.19}$$

There is a simple technique for eliminating the parameter v_1 from (6.18) and (6.19). First write each equation as a polynomial in v_1:

$$\begin{aligned} v_1^3 - 2(r_1 - c_1)v_1 + 4 &= 0, \\ v_1^3 - (s_1 - c_2)v_1^2 + 1 &= 0. \end{aligned}$$

Now eliminate the highest power of v_1 from one equation:

$$\begin{aligned} v_1^3 - 2(r_1 - c_1)v_1 + 4 &= 0, \\ (s_1 - c_2)v_1^2 - 2(r_1 - c_1)v_1 + 3 &= 0. \end{aligned}$$

Use the lower-order equation (multiplied by an appropriate power of v_1) to eliminate the highest power of v_1 from the other equation:

$$\begin{aligned} 2(r_1 - c_1)v_1^2 - \{2(r_1 - c_1)(s_1 - c_2) + 3\}v_1 + 4(s_1 - c_2) &= 0, \\ (s_1 - c_2)v_1^2 - 2(r_1 - c_1)v_1 + 3 &= 0. \end{aligned}$$

Iterate until v_1 is found; in this example, only one more elimination is needed to obtain a linear equation for v_1, whose solution is

$$v_1 = \frac{6(r_1 - c_1) - 4(s_1 - c_2)^2}{4(r_1 - c_1)^2 - 2(r_1 - c_1)(s_1 - c_2)^2 - 3(s_1 - c_2)} \tag{6.20}$$

(provided that the denominator is nonzero). The general solution of the ODE is obtained by substituting (6.20) into

$$(s_1 - c_2)v_1^2 - 2(r_1 - c_1)v_1 + 3 = 0$$

and replacing (r_1, s_1) by (y, x). Another way of reaching the solution is to continue using the elimination algorithm until v_1 vanishes from one equation.

The algorithm works for any parametric solution such that r and s are both rational polynomials in some parameter (which need not be v).

6.2 New Symmetries Obtained During Reduction

So far, we have assumed that the Lie algebra has a solvable subalgebra that is sufficiently large to enable us to solve the ODE completely. The remainder of this chapter deals with the problem of solving ODEs of order n whose largest solvable subalgebras are of dimension $n-1$ or less. Let $\{X_1, \ldots, X_S\}$ be a canonical basis for such a subalgebra. In terms of the fundamental differential invariants (r_S, v_S) of this subalgebra, the original ODE is equivalent to an ODE of order $n-S$. The general solution of the reduced ODE is an algebraic equation of the form

$$v_S = F(r_S; c_1, \ldots, c_{n-S}). \tag{6.21}$$

If this solution can be determined (perhaps by exploiting symmetries of the reduced ODE) the problem disappears, because (6.21) is equivalent to an ODE of order S that admits the symmetries generated by $\{X_1, \ldots, X_S\}$. What can be done if (6.21) cannot be found?

Each subalgebra $\mathcal{L}_k = \text{Span}(X_1, \ldots, X_k)$ in the solvable chain can be used to reduce the original ODE to an equivalent ODE of order $n-k$ in the fundamental differential invariants (r_k, v_k). Thus there is a sequence of "intermediate" reduced ODEs. Until now, we have used the whole of $\mathcal{L}_S$ in order to achieve the maximal reduction of order. This is of no use if the maximally reduced ODE cannot be solved. However, it may be that one of the intermediate ODEs has new point symmetries, as well as those inherited from the original ODE. With a sufficient number of new symmetries, it might be possible to obtain the general solution of an intermediate equation in the form

$$v_k = F(r_k; c_1, \ldots, c_{n-k}).$$

Then the symmetries in $\mathcal{L}_k$ can be used to complete the solution of the original ODE. The following example illustrates this idea.

Example 6.5 The third-order ODE

$$y''' = \frac{2y''^2}{y'} + \frac{y''}{x} + \frac{y'^2}{x} \tag{6.22}$$

has the two-dimensional abelian Lie algebra generated by

$$X_1 = \partial_y, \qquad X_2 = x\partial_x.$$

The fundamental differential invariants are

$$\begin{aligned} r_1 &= x, \qquad v_1 = y' \qquad \text{for} \quad \mathcal{L}_1 = \mathrm{Span}(X_1); \\ r_2 &= xy', \qquad v_2 = x^2 y'' \qquad \text{for} \quad \mathcal{L}_2 = \mathrm{Span}(X_1, X_2). \end{aligned} \tag{6.23}$$

Therefore the ODE (6.22) is equivalent to

$$\frac{dv_2}{dr_2} = \frac{2v_2^2 + 3r_2 v_2 + r_2^3}{r_2(v_2 + r_2)}, \tag{6.24}$$

whose symmetries are not obvious. However, (6.22) is also equivalent to the second-order ODE

$$\frac{d^2 v_1}{dr_1^2} = \frac{2}{v_1}\left(\frac{dv_1}{dr_1}\right)^2 + \frac{1}{r_1}\left(\frac{dv_1}{dr_1} + v_1^2\right), \tag{6.25}$$

which has an eight-dimensional Lie algebra of point symmetry generators. The symmetries generated by X_2 are inherited from the original ODE (6.22), but the remaining symmetries are new. Rather than examining the whole Lie algebra, we shall focus on the two-dimensional subalgebra $\tilde{\mathcal{L}}_2$ that is spanned by

$$\tilde{X}_1 = v_1^2 \partial_{v_1}, \qquad \tilde{X}_2 = r_1 \partial_{r_1} - v_1 \partial_{v_1}. \tag{6.26}$$

Note that $\tilde{X}_2$ is the restriction of the inherited generator X_2 to (r_1, v_1). The basis (6.26) is canonical, because

$$[\tilde{X}_1, \tilde{X}_2] = \tilde{X}_1.$$

Hence the ODE (6.25) can be reduced to quadratures, using the differential invariants

$$\begin{aligned} \tilde{r}_1 &= r_1, \qquad \tilde{v}_1 = \frac{1}{v_1^2}\frac{dv_1}{dr_1} \qquad \text{for} \quad \tilde{\mathcal{L}}_1 = \mathrm{Span}(\tilde{X}_1); \\ \tilde{r}_2 &= \frac{1}{v_1^2}\frac{dv_1}{dr_1}, \qquad \tilde{v}_2 = \frac{r_1}{v_1^2}\frac{d^2 v_1}{dr_1^2} - \frac{2r_1}{v_1^3}\left(\frac{dv_1}{dr_1}\right)^2 \qquad \text{for} \quad \tilde{\mathcal{L}}_2. \end{aligned} \tag{6.27}$$

Skipping the details, we arrive at the general solution of (6.25):

$$v_1 = \frac{2c_1}{(c_1 r_1 + 1)^2 + c_2}. \tag{6.28}$$

Finally, the canonical coordinates $(r_1, s_1) = (x, y)$ are used to complete the solution of (6.22), which is

$$y = \begin{cases} c_3 + 2\tilde{c}_2 \tan^{-1}\left(\tilde{c}_2(c_1 x + 1)\right), & c_2 = (\tilde{c}_2)^{-2} > 0; \\ c_3 - \dfrac{2}{c_1 x + 1}, & c_2 = 0; \\ c_3 + (\tilde{c}_2)^{-1} \ln \left| \dfrac{c_1 x + 1 - \tilde{c}_2}{c_1 x + 1 + \tilde{c}_2} \right|, & c_2 = -\tilde{c}_2^2 < 0. \end{cases} \tag{6.29}$$

This example shows that it is worthwhile calculating the point symmetries of the intermediate ODEs if the fully reduced ODE cannot be solved.

6.3 Integration of Third-Order ODEs with $\mathfrak{sl}(2)$

The three-dimensional Lie algebra $\mathfrak{sl}(2)$ is not solvable, and the above methods do not work for third-order ODEs with this Lie algebra. One of the simplest realizations of $\mathfrak{sl}(2)$ as a set of generators of Lie point symmetries is

$$X_1 = \partial_x, \qquad X_2 = x\partial_x, \qquad X_3 = x^2\partial_x. \tag{6.30}$$

The fundamental differential invariants for the Lie group generated by (6.30) are

$$r_\alpha = y, \qquad v_\alpha = \frac{2y'y''' - 3y''^2}{2y'^4}. \tag{6.31}$$

Hence every third-order ODE whose Lie point symmetries are generated by (6.30) is of the form

$$v_\alpha = F(r_\alpha) \tag{6.32}$$

for some function F. We can reduce (6.32) to a first-order ODE in the usual way, by using the solvable subgroup generated by X_1 and X_2. The reduced ODE is the Riccati equation

$$\frac{dz}{dy} + \tfrac{1}{2}z^2 = F(y), \qquad \text{where} \quad z = \frac{y''}{y'^2}, \tag{6.33}$$

which can be linearized by introducing the function

$$\psi(y) = \sqrt{y'}.$$

Then $z = 2\psi'(y)/\psi(y)$, where $\psi(y)$ satisfies the Schrödinger equation

$$\frac{d^2\psi}{dy^2} - \tfrac{1}{2}F(y)\psi = 0. \tag{6.34}$$

Suppose that $\psi(y)$ is an arbitrary nonzero solution of (6.34). If $\varphi(y)$ is any nonzero solution that is linearly independent of $\psi(y)$, the Wronskian

$$W = \psi(y)\varphi'(y) - \varphi(y)\psi'(y)$$

is a nonzero constant. Note that

$$\frac{d}{dy}\left(\frac{\varphi(y)}{\psi(y)}\right) = \frac{W}{\psi^2} = W\frac{dx}{dy},$$

and hence

$$Wx = \frac{\varphi(y)}{\psi(y)} + c.$$

For simplicity we rescale φ to give $W = 1$; this does not alter the solution. Moreover, we may take c to be zero (redefining φ if necessary). So the general solution of the third-order ODE (6.32) is

$$x = \frac{\varphi(y)}{\psi(y)}. \tag{6.35}$$

The solution depends upon three arbitrary constants, not four, because φ is normalized to ensure that $W = 1$.

The set of generators (6.30) is not the only realization of $\mathfrak{sl}(2)$ on the (x, y) plane. It is one of three distinct realizations that cannot be mapped to one another by any point transformation. The other two are

$$X_1 = \partial_x, \qquad X_2 = x\partial_x - y\partial_y, \qquad X_3 = x^2\partial_x - 2xy\partial_y. \tag{6.36}$$

$$X_1 = \partial_x, \qquad X_2 = x\partial_x - y\partial_y, \qquad X_3 = x^2\partial_x - (2xy+1)\partial_y. \tag{6.37}$$

Each of the three realizations is a representative of an equivalence class of realizations that can be mapped to one another by a complex point transformation.

For any ODE whose generators can be mapped to (6.30) by a point transformation, the solution strategy is obvious: carry out the transformation, solve the Schrödinger equation (6.34) (if possible), and then transform the solution back to the original variables. Remarkably, the same strategy works for ODEs whose generators are equivalent to one of the other two realizations: the only difference is that the required transformation is not a point transformation.

The first prolongation of the set of generators (6.30) is

$$X_1 = \partial_x, \qquad X_2 = x\partial_x - y'\partial_{y'}, \qquad X_3 = x^2\partial_x - 2xy'\partial_{y'}.$$

Let $p = y'$; then the prolonged generators are

$$X_1 = \partial_x, \qquad X_2 = x\partial_x - p\partial_p, \qquad X_3 = x^2\partial_x - 2xp\partial_p, \tag{6.38}$$

which is the second realization (6.36) (with p replacing y). The most general third-order ODE with the symmetries generated by (6.38) is

$$v_\beta = G(r_\beta), \tag{6.39}$$

where the fundamental differential invariants are

$$r_\beta = v_\alpha = \frac{2pp'' - 3p'^2}{2p^4}, \qquad v_\beta = \frac{dv_\alpha}{dr_\alpha} = \frac{p^2p''' - 6pp'p'' + 6p'^3}{p^6}. \tag{6.40}$$

Therefore

$$\int \frac{dv_\alpha}{G(v_\alpha)} = r_\alpha + c = y + c. \tag{6.41}$$

Our aim is to determine $p = y'$ as a function of x, so we may set c to any convenient value, without affecting the result. Provided that (6.41) can be solved to obtain an expression of

$$v_\alpha = F(r_\alpha),$$

we can reduce the problem to the one treated earlier. Once the solution

$$x = \frac{\varphi(y)}{\psi(y)} \tag{6.42}$$

has been found, it is easy to obtain

$$p = \frac{dy}{dx} = \left(\psi(y)\right)^2. \tag{6.43}$$

Taken together, (6.42) and (6.43) constitute a parametric solution of the ODE (6.39).

Example 6.6 To illustrate the method, we shall complete the solution of the fourth-order ODE

$$\tilde{y}^{(iv)} = \frac{2}{\tilde{y}}(1 - \tilde{y}')\tilde{y}'''. \tag{6.44}$$

(Here $\tilde{y}$ is used in place of y to avoid confusion later.) In Example 5.2, we reduced this ODE to

$$\tilde{y}''' = \frac{2\tilde{y}\tilde{y}'' - \tilde{y}'^2 + c_1}{\tilde{y}^2}, \tag{6.45}$$

whose Lie point symmetry generators are

$$X_1 = \partial_x, \qquad X_2 = x\partial_x + \tilde{y}\partial_{\tilde{y}}, \qquad X_3 = x^2\partial_x + 2x\tilde{y}\partial_{\tilde{y}}. \tag{6.46}$$

Let $p = 1/\tilde{y}$; then the generators (6.46) are equivalent to (6.36). The reduced ODE (6.45) is equivalent to

$$v_\beta = 2r_\beta - c_1.$$

Therefore

$$v_\alpha = F(r_\alpha) \equiv \tfrac{1}{2}c_1 + \exp\{2(r_\alpha + c)\}.$$

It is convenient to choose $c = \frac{1}{2}(\ln 2 + \pi i)$, so that the Schrödinger equation is

$$\frac{d^2\psi}{dy^2} + \left(e^{2y} - \tfrac{1}{4}c_1\right)\psi = 0.$$

This is equivalent to Bessel's equation

$$t^2\frac{d^2\psi}{dt^2} + t\frac{d\psi}{dt} + (t^2 - \nu^2)\psi = 0, \qquad t = e^y, \quad \nu = \tfrac{1}{2}\sqrt{c_1}. \tag{6.47}$$

So, from (6.42) and (6.43), the general solution of the third-order ODE (6.45) is

$$\begin{aligned} x &= \frac{c_2 J_\nu(t) + c_3 Y_\nu(t)}{c_4 J_\nu(t) + c_5 Y_\nu(t)}, \\ p &= \left(c_4 J_\nu(t) + c_5 Y_\nu(t)\right)^2, \end{aligned} \tag{6.48}$$

where $J_\nu(t)$ and $Y_\nu(t)$ are Bessel functions and either c_2 or c_3 is chosen so that

$$W = \frac{2}{\pi}(c_3c_4 - c_2c_5) = 1.$$

Finally, we obtain the solution of (6.44) from the identity $\tilde{y} = p^{-1}$.

The third realization, (6.37), is related to the second realization in much the same way as the second is related to the first. Prolonging the generators (6.38) once, we obtain

$$\begin{aligned} X_1 &= \partial_x, \qquad X_2 = x\partial_x - p\partial_p - 2p'\partial_{p'}, \\ X_3 &= x^2\partial_x - 2xp\partial_p - (4xp' + 2p)\partial_{p'}, \end{aligned}$$

which is equivalent to

$$X_1 = \partial_x, \qquad X_2 = x\partial_x - p\partial_p - q\partial_q,$$
$$X_3 = x^2\partial_x - 2xp\partial_p - (2xq+1)\partial_q,$$

where

$$q(x) = \frac{p'(x)}{2p(x)}. \tag{6.49}$$

Therefore the restriction of prolonged second realization to functions of x and q is

$$X_1 = \partial_x, \qquad X_2 = x\partial_x - q\partial_q, \qquad X_3 = x^2\partial_x - (2xq+1)\partial_q, \tag{6.50}$$

which is equivalent to the third realization. The fundamental differential invariants of (6.50) are

$$\begin{aligned} r_\gamma &= \frac{v_\beta}{r_\beta^{3/2}} = \frac{q'' - 6qq' + 4q^3}{\sqrt{2}\,(q'-q^2)^{3/2}}, \\ v_\gamma &= \frac{2v_\beta}{r_\beta^2}\frac{dv_\beta}{dr_\beta} = \frac{q''' - 12qq'' + 18q'^2}{(q'-q^2)^2} - 24. \end{aligned} \tag{6.51}$$

Hence the general third-order ODE whose Lie point symmetries are generated by (6.50), namely

$$v_\gamma = H(r_\gamma),$$

is equivalent to

$$\frac{2v_\beta}{r_\beta^2}\frac{dv_\beta}{dr_\beta} = H\left(\frac{v_\beta}{r_\beta^{3/2}}\right). \tag{6.52}$$

This first-order ODE has scaling symmetries (which are generated by $X = 2r_\beta\partial_{r_\beta} + 3v_\beta\partial_{v_\beta}$); these enable us to reduce the ODE to quadrature:

$$r_\beta = c\exp\left\{\int \frac{2r_\gamma\, dr_\gamma}{H(r_\gamma) - 3r_\gamma^2}\right\}. \tag{6.53}$$

Suppose that (6.53) can be rearranged to give r_γ in terms of r_β. Then the problem is reduced to that of finding the general solution of an ODE whose Lie point symmetries belong to the second realization, namely

$$v_\beta = G(r_\beta) \equiv r_\beta^{3/2} r_\gamma(r_\beta). \tag{6.54}$$

We already know from (6.42) and (6.43) that the general solution of (6.54) is of the form

$$x = \frac{\varphi(y)}{\psi(y)}, \qquad p = \frac{dy}{dx} = (\psi(y))^2.$$

Therefore

$$q = \frac{p'(x)}{2p(x)} = \psi(y)\psi'(y).$$

So the parametric solution of a general ODE whose symmetries are in the third realization is

$$x = \frac{\varphi(y)}{\psi(y)}, \qquad q = \psi(y)\psi'(y). \tag{6.55}$$

It is remarkable that all three realizations are related by prolongation so that (in principle) everything reduces to a study of the Schrödinger equation. However, the method also requires that the quadratures at each stage should be tractable.

Further Reading

The method of successive integration described in §6.1 is presented in various forms in the literature. Stephani (1989) describes several different versions, including Lie's algorithm for constructing line integrals at each stage.

For most of this book, we ignore the problem of the existence of closed-form solutions. Cox, Little, and O'Shea (1992) includes an easy introduction to implicitization and various related problems.

Section 6.3 is based on a seminal paper by Clarkson and Olver (1996), which particularly focuses on the Chazy equation. This paper clearly illustrates the value of symmetry methods for dealing with difficult analytical problems.

Exercises

6.1 Find the general solution of

$$y'' = y'(1 - y')/y,$$

whose Lie point symmetries are generated by $X_1 = \partial_x$, $X_2 = x\partial_x + y\partial_y$.

6.2 Solve the ODE

$$y'' = \frac{yy'}{x^3} - y^2x^4.$$

6.3 Derive (6.28).

6.4 Solve the ODE

$$y'' = \frac{1}{xy^2} .$$

6.5 The ODE

$$y'' = \frac{y'^2}{y} - \frac{y^2}{x^3 y'}$$

has a two-dimensional Lie algebra that is spanned by $X_1 = x\partial_x$, $X_2 = y\partial_y$. Use these generators, in a suitable order, to solve the ODE.

6.6 Solve the ODE

$$y''' = 3y''^2/(2y') + \left(\tfrac{1}{2}y^2 + 1\right)y'^3,$$

which has the $\mathfrak{sl}(2)$ Lie algebra spanned by

$$X_1 = \partial_x, \qquad X_2 = x\partial_x, \qquad X_3 = x^2\partial_x .$$

6.7 Use the symmetry generators $X_1 = \partial_x, X_2 = x\partial_x + y\partial_y$ to obtain the general solution to

$$y'' = \frac{y' - 1}{y}$$

in parametric form. Now eliminate the parameter to obtain the solution in the form

$$y = F(x; c_1, c_2).$$

7

Techniques Based on First Integrals

The junior Bat asked the senior Bat
A question most profound:
'How do the humans down below
Hang by their feet from the ground?'

(Dick Smithells and Ian Pillinger: Alphabet Zoop)

7.1 First Integrals Derived from Symmetries

Second-order ODEs whose Lie algebra is $\mathfrak{so}(3)$ cannot be solved by using a two-dimensional solvable subalgebra, for no such subalgebra exists. However these ODEs can be solved by a different approach. In this section, we derive a simple way of using Lie point symmetries to determine *first integrals* of a given ODE. This technique relies upon the dimension of the Lie algebra being higher than the order of the ODE. [Recall: $\mathfrak{so}(3)$ is three-dimensional.] Remarkably, the method enables us to solve some ODEs without having to carry out any quadrature whatsoever!

A first integral of the ODE

$$y^{(n)} = \omega\big(x, y, y', \dots, y^{(n-1)}\big) \tag{7.1}$$

is a nonconstant function

$$\phi\big(x, y, y', \dots, y^{(n-1)}\big), \tag{7.2}$$

that is constant on solutions of the ODE. Hence

$$D_x\phi = 0 \qquad \text{when (7.1) holds.} \tag{7.3}$$

A neater form of the defining equation is

$$\bar{D}\phi = 0, \qquad \phi_{y^{(n-1)}} \neq 0, \tag{7.4}$$

where

$$\bar{D} = \partial_x + y'\partial_y + \cdots + y^{(n-1)}\partial_{y^{(n-2)}} + \omega\partial_{y^{(n-1)}}. \tag{7.5}$$

So far, we have generally characterized Lie point symmetries by their infinitesimal generators, but throughout this chapter it will be more convenient to work in terms of the characteristic, $Q = \eta - y'\xi$. The linearized symmetry condition can be expressed in terms of Q as follows:

$$\bar{D}^n Q - \omega_{y^{(n-1)}}\bar{D}^{n-1} Q - \cdots - \omega_{y'}\bar{D}Q - \omega_y Q = 0. \tag{7.6}$$

To derive this result, we substitute the identities

$$\eta^{(k)} = D_x^k Q + y^{(k+1)}\xi, \qquad k = 0, \ldots, n, \tag{7.7}$$

into the linearized symmetry condition (3.12), obtaining

$$\begin{aligned} D_x^n Q - \omega_{y^{(n-1)}} D_x^{n-1} Q - \cdots - \omega_{y'} D_x Q - \omega_y Q & \\ + \xi D_x\left(y^{(n)} - \omega\right) = 0 \qquad \text{when} \quad y^{(n)} = \omega. & \end{aligned} \tag{7.8}$$

The terms that are multiplied by ξ vanish, because every solution of the ODE (7.1) is a solution of

$$D_x^k\left(y^{(n)} - \omega\right) = 0, \qquad k = 0, 1, 2, \ldots. \tag{7.9}$$

Now we isolate terms of the form $\bar{D}^k Q$ in each derivative; for example,

$$D_x Q = \bar{D}Q + HQ,$$

where

$$H = \left(y^{(n)} - \omega\right)\partial_{y^{(n-1)}}. \tag{7.10}$$

Similarly

$$D_x^2 Q = \bar{D}^2 Q + H(\bar{D}Q) + D_x(HQ),$$

and, more generally,

$$D_x^k Q = \bar{D}^k Q + \sum_{j=0}^{k-1} D_x^{k-1-j}\left(H(\bar{D}^j Q)\right). \tag{7.11}$$

Taking (7.9) into account, (7.11) yields

$$D_x^k Q\big|_{y^{(n)}=\omega} = \bar{D}^k Q, \qquad k = 0, 1, 2, \ldots$$

for any function $Q(x, y, y', \ldots, y^{(n-1)})$, and thus (7.8) is equivalent to (7.6).

Suppose that the infinitesimal generators $X_1, \ldots, X_R$ constitute a basis for the Lie algebra of point symmetry generators of a given ODE (7.1). Then the corresponding characteristics $Q_1, \ldots, Q_R$ form a basis for the set of all solutions of (7.6) that depend only upon (x, y, y') and are linear in y'. Other solutions of (7.6) are discussed in the next section.

For now, let us restrict our attention to second-order ODEs

$$y'' = \omega(x, y, y') \tag{7.12}$$

whose Lie algebra $\mathcal{L}$ has dimension $R > 2$. A function $\phi(x, y, y')$ is a first integral of (7.12) if

$$\bar{D}\phi = 0, \qquad \phi_{y'} \neq 0, \tag{7.13}$$

where

$$\bar{D} = \partial_x + y'\partial_y + \omega\partial_{y'}. \tag{7.14}$$

If we can find two functionally independent first integrals ϕ^1 and ϕ^2, we obtain the general solution of the ODE in the parametric form

$$\phi^1(x, y, y') = c_1, \qquad \phi^2(x, y, y') = c_2. \tag{7.15}$$

Here y' acts as a parameter, which must be eliminated if the general solution is to be written in closed form.

Given a basis $\{X_1, \ldots, X_R\}$ for $\mathcal{L}$, we use Q_i to denote the characteristic corresponding to X_i. Let

$$W_{ij} = Q_i\bar{D}Q_j - Q_j\bar{D}Q_i, \qquad 1 \leq i < j \leq R. \tag{7.16}$$

The linearized symmetry condition (7.6) is

$$\bar{D}^2 Q_i - \omega_{y'}\bar{D}Q_i - \omega_y Q_i = 0. \tag{7.17}$$

and hence

$$\bar{D}W_{ij} = Q_i\bar{D}^2 Q_j - Q_j\bar{D}^2 Q_i = \omega_{y'} W_{ij}. \tag{7.18}$$

The ratio of any two nonzero W_{ij} is either a constant or else a first integral, because (7.18) yields

$$\bar{D}\left(\frac{W_{ij}}{W_{kl}}\right) = 0.$$

First integrals also arise when $W_{ij} = 0$, for then

$$\bar{D}\left(\frac{Q_j}{Q_i}\right) = 0.$$

[The ratio Q_j/Q_i is a first integral, not a constant, because X_i and X_j are linearly independent.] With these results, it is straightforward to calculate first integrals.

N.B. For some functions ω, an extra first integral may be obtained from (7.18). For example, if $\omega_{y'} = 0$ then every nonconstant W_{ij} is a first integral.

Example 7.1 The Lie algebra $\mathfrak{so}(3)$ is realized by the symmetry generators

$$\begin{gathered} X_1 = y\partial_x - x\partial_y, \qquad X_2 = \tfrac{1}{2}(1 + x^2 - y^2)\partial_x + xy\partial_y, \\ X_3 = xy\partial_x + \tfrac{1}{2}(1 - x^2 + y^2)\partial_y. \end{gathered} \tag{7.19}$$

These generators have the characteristics

$$\begin{gathered} Q_1 = -x - yy', \qquad Q_2 = xy - \tfrac{1}{2}(1 + x^2 - y^2)y', \\ Q_3 = \tfrac{1}{2}(1 - x^2 + y^2) - xyy'. \end{gathered} \tag{7.20}$$

We shall use the method described above to solve the ODE

$$y'' = \frac{2(xy' - y)(1 + y'^2)}{1 + x^2 + y^2}, \tag{7.21}$$

which has the Lie point symmetries generated by (7.19). First, we apply the operator

$$\bar{D} = \partial_x + y'\partial_y + \omega\partial_{y'}, \qquad \text{where} \quad \omega = \frac{2(xy' - y)(1 + y'^2)}{1 + x^2 + y^2},$$

to each characteristic in turn, obtaining

$$\begin{aligned} \bar{D}Q_1 &= -(1 + y'^2) - y\omega, \\ \bar{D}Q_2 &= y(1 + y'^2) + \tfrac{1}{2}(y^2 - x^2 - 1)\omega, \\ \bar{D}Q_3 &= -x(1 + y'^2) - xy\omega. \end{aligned} \tag{7.22}$$

Now we calculate each W_{ij}, substituting the right-hand side of (7.21) for ω:

$$\begin{aligned}
W_{12} &= \tfrac{1}{2}(1+y'^2)\{2x(xy'-y) - y'(1+x^2+y^2)\},\\
W_{13} &= \tfrac{1}{2}(1+y'^2)\{2y(xy'-y) + 1 + x^2 + y^2\},\\
W_{23} &= (1+y'^2)(xy'-y).
\end{aligned} \tag{7.23}$$

Therefore two functionally independent first integrals of (7.21) are

$$\begin{aligned}
\phi^1 &= \frac{W_{12}}{W_{23}} = x - \frac{y'(1+x^2+y^2)}{2(xy'-y)},\\
\phi^2 &= \frac{W_{13}}{W_{23}} = y + \frac{1+x^2+y^2}{2(xy'-y)}.
\end{aligned} \tag{7.24}$$

Eliminating y' from (7.15), we obtain the general solution

$$(x-c_1)^2 + (y-c_2)^2 = 1 + c_1^2 + c_2^2. \tag{7.25}$$

(N.B. There are also solutions of the form $y = cx$, on which W_{23} vanishes.) Every second-order ODE whose Lie algebra is $\mathfrak{so}(3)$ may be solved in this way.

To see how the above method is derived, and how it may be generalized, consider the third-order ODE

$$y''' = \omega(x, y, y', y''), \tag{7.26}$$

The linearized symmetry condition is

$$\bar{D}^3 Q - \omega_{y''}\bar{D}^2 Q - \omega_{y'}\bar{D}Q - \omega_y Q = 0, \tag{7.27}$$

where now

$$\bar{D} = \partial_x + y'\partial_y + y''\partial_{y'} + \omega\partial_{y''}. \tag{7.28}$$

First integrals are the solutions of

$$\bar{D}\phi = 0, \qquad \phi_{y''} \neq 0.$$

Suppose that $R \geq 4$, and consider any four linearly independent characteristic functions $Q_1, \ldots, Q_4$ satisfying (7.27). The rank of the matrix

$$\mathcal{Q}_{1234} = \begin{bmatrix} Q_1 & Q_2 & Q_3 & Q_4 \\ \bar{D}Q_1 & \bar{D}Q_2 & \bar{D}Q_3 & \bar{D}Q_4 \\ \bar{D}^2Q_1 & \bar{D}^2Q_2 & \bar{D}^2Q_3 & \bar{D}^2Q_4 \\ \bar{D}^3Q_1 & \bar{D}^3Q_2 & \bar{D}^3Q_3 & \bar{D}^3Q_4 \end{bmatrix}$$

is three or less, because (7.27) enables us to write the fourth row of $\mathcal{Q}_{1234}$ as a linear combination of the other three rows. Let

$$\mathcal{Q}_{ijk} = \begin{bmatrix} Q_i & Q_j & Q_k \\ \bar{D}Q_i & \bar{D}Q_j & \bar{D}Q_k \\ \bar{D}^2Q_i & \bar{D}^2Q_j & \bar{D}^2Q_k \end{bmatrix},$$

and let

$$W_{ijk} = \det(\mathcal{Q}_{ijk}).$$

Note that

$$\bar{D}W_{ijk} = \omega_{y''} W_{ijk}. \tag{7.29}$$

If the rank of $\mathcal{Q}_{1234}$ is three then, without loss of generality, we may suppose that the column space of $\mathcal{Q}_{1234}$ is spanned by the first three columns (renumbering the functions Q_i if necessary). Then there exist functions μ_i such that

$$\mathcal{Q}_{123} \begin{bmatrix} \mu_1 \\ \mu_2 \\ \mu_3 \end{bmatrix} = \begin{bmatrix} Q_4 \\ \bar{D}Q_4 \\ \bar{D}^2Q_4 \end{bmatrix} \tag{7.30}$$

and

$$\mu_1\bar{D}^3Q_1 + \mu_2\bar{D}^3Q_2 + \mu_3\bar{D}^3Q_3 = \bar{D}^3Q_4. \tag{7.31}$$

By applying the operator $\bar{D}$ to (7.30) and then taking (7.30) and (7.31) into account, we obtain

$$\mathcal{Q}_{123} \begin{bmatrix} \bar{D}\mu_1 \\ \bar{D}\mu_2 \\ \bar{D}\mu_3 \end{bmatrix} = \begin{bmatrix} 0 \\ 0 \\ 0 \end{bmatrix}. \tag{7.32}$$

We have assumed that Q_{123} is nonsingular, and therefore

$$\bar{D}\mu_i = 0, \qquad i = 1, 2, 3. \tag{7.33}$$

Hence the functions μ_i are either first integrals or constants. To find out which, we solve (7.30) using Cramer's rule:

$$\mu_1 = \frac{W_{423}}{W_{123}}, \qquad \mu_2 = \frac{W_{143}}{W_{123}}, \qquad \mu_3 = \frac{W_{124}}{W_{123}}. \tag{7.34}$$

If (7.34) yields only two functionally independent first integrals, it is worth checking to see whether an extra first integral is obtainable from the relationship (7.29).

So far, we have assumed that $\mathrm{rank}(\mathcal{Q}_{1234}) = 3$, so that at least one of the determinants W_{ijk} is nonzero. If $\mathrm{rank}(\mathcal{Q}_{1234}) = 2$, we may apply the above argument to each matrix $\mathcal{Q}_{ijk}$ that is of rank two. Suppose that $\mathrm{rank}(\mathcal{Q}_{123}) = 2$, and that the column space of $\mathcal{Q}_{123}$ is spanned by the first two columns. Let

$$\mathcal{Q}_{ij} = \begin{bmatrix} Q_i & Q_j \\ \bar{D}Q_i & \bar{D}Q_j \end{bmatrix},$$

and let

$$W_{ij} = \det(\mathcal{Q}_{ij}) = Q_i \bar{D} Q_j - Q_j \bar{D} Q_i.$$

Repeating the argument that led to (7.33) and (7.34), we obtain

$$\bar{D}\left(\frac{W_{13}}{W_{12}}\right) = 0, \qquad \bar{D}\left(\frac{W_{23}}{W_{12}}\right) = 0. \tag{7.35}$$

Hence the ratios of the functions W_{ij} are either first integrals or constants.

To summarize, the strategy for finding first integrals of third-order ODEs (with four symmetry generators) is as follows. If $\mathrm{rank}(\mathcal{Q}_{1234}) = 3$, calculate the ratios of any nonzero determinants W_{ijk}. (There is no need to consider ratios of subdeterminants of any $\mathcal{Q}_{ijk}$ for which $W_{ijk} = 0$, as these are already included in the ratios that have been calculated.) If this yields only two first integrals, check to see whether an extra one can be found from (7.29). If $\mathrm{rank}(\mathcal{Q}_{1234}) = 2$ then one should calculate the ratios of any nonzero determinants W_{ij}.

The generalization of the above method to higher-order ODEs is obvious. Given an ODE of order n with $R \geq n + 1$ generators of Lie point symmetries, begin by calculating

$$\nu = \mathrm{rank}\begin{bmatrix} Q_1 & Q_2 & \cdots & Q_R \\ \bar{D}Q_1 & \bar{D}Q_2 & \cdots & \bar{D}Q_R \\ \vdots & \vdots & & \vdots \\ \bar{D}^{n-1}Q_1 & \bar{D}^{n-1}Q_2 & \cdots & \bar{D}^{n-1}Q_R \end{bmatrix}. \tag{7.36}$$

Then calculate determinants of $\nu \times \nu$ matrices of the form

$$\mathcal{Q}_{i_1\ldots i_\nu} = \begin{bmatrix} Q_{i_1} & Q_{i_2} & \cdots & Q_{i_\nu} \\ \bar{D}Q_{i_1} & \bar{D}Q_{i_2} & \cdots & \bar{D}Q_{i_\nu} \\ \vdots & \vdots & & \vdots \\ \bar{D}^{\nu-1}Q_{i_1} & \bar{D}^{\nu-1}Q_{i_2} & \cdots & \bar{D}^{\nu-1}Q_{i_\nu} \end{bmatrix}. \tag{7.37}$$

The ratios of the nonzero determinants are either first integrals or constants. If $\nu = n$ and the method yields only $n-1$ first integrals, try to obtain an extra one from

$$\bar{D}W_{i_1\ldots i_\nu} = \omega_{y^{(n-1)}} W_{i_1\ldots i_\nu}, \qquad \text{where} \quad W_{i_1\ldots i_\nu} = \det(\mathcal{Q}_{i_1\ldots i_\nu}). \tag{7.38}$$

For point symmetries, each Q_i is a function of x, y, and y' only. Therefore

$$\partial_{y^{(n-1)}}\left(\bar{D}^{k}Q_i\right) = 0, \qquad k = 1, \ldots, n-3.$$

All first integrals involve $y^{(n-1)}$, and hence we can obtain first integrals by the above procedure only if $\nu \in \{n-1, n\}$. Usually $\nu = n$, but the other possibility may occur if the full Lie algebra is not used.

Example 7.2 To illustrate the method described above, consider the ODE

$$y''' = \frac{(3y'-1)y''^2}{y'^2}, \tag{7.39}$$

which has the four-dimensional Lie algebra spanned by

$$X_1 = \partial_y, \qquad X_2 = \partial_x, \qquad X_3 = x\partial_x + y\partial_y, \qquad X_4 = y\partial_x. \tag{7.40}$$

Therefore

$$\nu = \text{rank} \begin{bmatrix} 1 & -y' & y - xy' & -yy' \\ 0 & -y'' & -xy'' & -yy'' - y'^2 \\ 0 & -\omega & -x\omega - y'' & -y\omega - 3y'y'' \end{bmatrix} = 3 \quad (\text{if } y'' \neq 0).$$

(As usual, ω denotes the right-hand side of the ODE.) Now we calculate the determinants W_{ijk}:

$$W_{123} = y''^2,$$
$$W_{124} = 3y'y''^2 - y'^2\omega = y''^2,$$

$$W_{134} = (3xy' - y)y''^2 - y'^2y'' - xy'^2\omega = (x - y)y''^2 - y'^2y'',$$
$$W_{234} = -3yy'y''^2 + y'^3y'' + yy'^2\omega = -yy''^2 + y'^3y''.$$

Because $W_{123} = W_{124}$, we obtain only two first integrals:

$$\phi^1 = x - y - \frac{y'^2}{y''},$$
$$\phi^2 = -y + \frac{y'^3}{y''}. \tag{7.41}$$

However, the relationship

$$\bar{D}W_{ijk} = \omega_{y''}W_{ijk} = \frac{2(3y' - 1)y''}{y'^2}W_{ijk}$$

(with $W_{ijk} = W_{123}$) leads to the third (functionally independent) first integral

$$\phi^3 = 2\ln|y''| - 6\ln|y'| - \frac{2}{y'}. \tag{7.42}$$

Eliminating y' and y'' from the equations $\phi^i = c_i$, we obtain the general solution

$$x = y + c_1 - (y + c_2)\left(\tfrac{1}{2}c_3 + \ln|y + c_2|\right). \tag{7.43}$$

This method is quite easy to use, particularly with the assistance of a reliable computer algebra system. It succeeds in solving second-order ODEs whose Lie algebra is $\mathfrak{so}(3)$. For other ODEs, it can provide a useful shortcut to the general solution. The method may be used with any sufficiently large set of linearly independent solutions of

$$\bar{D}^nQ - \omega_{y^{(n-1)}}\bar{D}^{n-1}Q - \cdots - \omega_{y'}\bar{D}Q - \omega_yQ = 0. \tag{7.44}$$

The linearized symmetry condition (7.44) has infinitely many solutions, most of which are not characteristics of Lie point symmetries. Instead, they are associated with higher-order symmetries.

7.2 Contact Symmetries and Dynamical Symmetries

So far, we have looked only for symmetries that are point transformations, that is, diffeomorphisms of the plane. Every generator of a one-parameter Lie group of point transformations has a characteristic $Q(x, y, y')$ that is linear in y'; the functions ξ, η and $\eta^{(1)}$ are expressible in terms of Q and its first derivatives as follows:

$$\xi = -Q_{y'}, \qquad \eta = Q - y'Q_{y'}, \qquad \eta^{(1)} = Q_x + y'Q_y. \tag{7.45}$$

A *contact transformation* is a diffeomorphism of the variables x, y, and y' that extends to derivatives by the usual prolongation conditions

$$\hat{y}^{(k+1)} = \frac{d\hat{y}^{(k)}}{d\hat{x}}, \qquad k = 0, 1, \ldots. \tag{7.46}$$

Therefore a diffeomorphism

$$(\hat{x}, \hat{y}, \hat{y}') = \big(\hat{x}(x, y, y'), \hat{y}(x, y, y'), \hat{y}'(x, y, y')\big) \tag{7.47}$$

is a contact transformation if (7.46) is satisfied for $k = 0$, that is, if

$$\hat{y}'(x, y, y') = \frac{\hat{y}_x + y'\hat{y}_y + y''\hat{y}_{y'}}{\hat{x}_x + y'\hat{x}_y + y''\hat{x}_{y'}}. \tag{7.48}$$

As $\hat{y}'$ is required to be independent of y'', every contact transformation (7.47) satisfies the *contact condition*

$$\hat{y}_{y'} = \hat{y}'\hat{x}_{y'}. \tag{7.49}$$

Consequently, (7.48) leads to

$$\hat{y}_x + y'\hat{y}_y = \hat{y}'(\hat{x}_x + y'\hat{x}_y). \tag{7.50}$$

One-parameter Lie groups of contact transformations are constructed by expanding about the identity, as follows:

$$\begin{aligned} \hat{x} &= x + \varepsilon\xi(x, y, y') + O(\varepsilon^2), \\ \hat{y} &= y + \varepsilon\eta(x, y, y') + O(\varepsilon^2), \\ \hat{y}' &= y' + \varepsilon\eta^{(1)}(x, y, y') + O(\varepsilon^2), \\ \hat{y}^{(k)} &= y^{(k)} + \varepsilon\eta^{(k)}\big(x, y, y', \ldots, y^{(k)}\big) + O(\varepsilon^2). \end{aligned} \tag{7.51}$$

Just as for Lie point transformations, each $\eta^{(k)}$ depends upon derivatives of order k or smaller. However, ξ and η may now depend upon x, y, and y', and so $Q(x, y, y') = \eta - y'\xi$ need not be linear in y'. The prolongation rules are the same as for Lie point transformations:

$$\eta^{(k)} = D_x^k Q + y^{(k+1)}\xi. \tag{7.52}$$

In particular,

$$\eta^{(1)} = Q_x + y'Q_y + y''(Q_{y'} + \xi),$$

so $\eta^{(1)}$ is independent of y'' if and only if $\xi = -Q_{y'}$. Hence the relationships (7.45) hold for all Lie contact transformations, not just those that are point transformations.

A *Lie contact symmetry* is a one-parameter Lie group of contact transformations whose characteristic satisfies the symmetry condition (7.44). Every Lie point symmetry is a Lie contact symmetry. Second-order ODEs have an infinite set of Lie contact symmetries (just as first-order ODEs have infinitely many Lie point symmetries), but it is not usually possible to find any. However, the Lie contact symmetries of a given ODE of order $n \geq 3$ can usually be found by splitting the symmetry condition, using the fact that Q is independent of $y'', \ldots, y^{(n-1)}$. This is essentially the same technique as is used to find Lie point symmetries.

Example 7.3 In Chapter 3, it was shown that the simplest third-order ODE,

$$y''' = 0, \tag{7.53}$$

has a seven-dimensional Lie algebra of Lie point symmetry generators. The characteristic corresponding to any such symmetry is a linear combination of

$$\begin{aligned} &Q_1 = 1, \qquad Q_2 = x, \qquad Q_3 = x^2, \qquad Q_4 = y, \\ &Q_5 = -y', \qquad Q_6 = -xy', \qquad Q_7 = 2xy - x^2y'. \end{aligned} \tag{7.54}$$

However, the ODE (7.53) also has nonpoint Lie contact symmetries. These are found by substituting $Q = Q(x, y, y')$ into (7.44) to obtain

$$\begin{aligned} 0 = &\left(Q_{xxx} + 3y'Q_{xxy} + 3y'^2Q_{xyy} + y'^3Q_{yyy}\right) \\ &+ 3y''\left(Q_{xxy'} + 2y'Q_{xyy'} + y'^2Q_{yyy'} + Q_{xy} + y'Q_{yy}\right) \\ &+ 3y''^2\left(Q_{xy'y'} + y'Q_{yy'y'} + Q_{yy'}\right) + y''^3\left(Q_{y'y'y'}\right). \end{aligned} \tag{7.55}$$

As Q is independent of y'', it follows that each set of terms enclosed by parentheses vanishes. The y''^3 terms vanish if

$$Q = A(x, y)y'^2 + B(x, y)y' + C(x, y)$$

for some functions A, B, and C. Then the y''^2 terms yield

$$(2A_x + B_y) + 4y'(A_y) = 0.$$

Equating powers of y', we obtain

$$A = \alpha(x), \qquad B = -2\alpha'(x)y + \beta(x)$$

for some functions α and β. Continuing in this way, we eventually obtain the general solution of (7.55), which depends upon ten arbitrary constants. Besides the seven characteristics listed in (7.54), there are three characteristics corresponding to nonpoint contact symmetries:

$$Q_8 = -y'^2, \qquad Q_9 = 2yy' - xy'^2, \qquad Q_{10} = (2y - xy')^2. \tag{7.56}$$

The infinitesimal generators of these nonpoint contact symmetries are

$$\begin{gathered} X_8 = 2y'\partial_x + y'^2\partial_y, \qquad X_9 = 2(xy' - y)\partial_x + xy'^2\partial_y + y'^2\partial_{y'}, \\ X_{10} = 2x(2y - xy')\partial_x + (4y^2 - x^2y'^2)\partial_y + 2y'(2y - xy')\partial_{y'}. \end{gathered} \tag{7.57}$$

After finding the set of all $Q(x, y, y')$ corresponding to contact symmetries of a given ODE, one can determine first integrals by using the method outlined in §7.1 (if there are sufficiently many symmetries).

Example 7.4 Consider the third-order ODE

$$y''' = x(x-1)y''^3 - 2xy''^2 + y'', \tag{7.58}$$

whose Lie contact symmetry generators have the characteristics

$$\begin{gathered} Q_1 = 1, \qquad Q_2 = x, \qquad Q_3 = e^{y'}, \\ Q_4 = (xy' - y - x)e^{y'}, \qquad Q_5 = e^{y - xy' + y' + x}. \end{gathered} \tag{7.59}$$

There are two Lie point symmetry generators, whose characteristics are Q_1 and Q_2, but these are not sufficient to solve the ODE. [If one writes (7.58) in terms of the differential invariants $(r_2, v_2) = (x, y'')$, the reduced ODE is

$$v_2' = r_2(r_2 - 1)v_2^3 - 2r_2v_2^2 + v_2,$$

whose solution is not obvious.] Nevertheless, the ODE (7.58) can be solved by using the contact symmetries to construct first integrals. The method of §7.1

yields the three functionally independent first integrals

$$\begin{aligned}\phi^1 &= \frac{W_{124}}{W_{123}} = xy' - y - \frac{1}{(x-1)y''-1},\\ \phi^2 &= \frac{W_{125}}{W_{123}} = \left(1 - x + \frac{1}{y''}\right)e^{y-xy'+x},\\ \phi^3 &= \frac{W_{134}}{W_{123}} = \left(1 + \frac{y''}{(x-1)y''-1}\right)e^{y'}.\end{aligned} \tag{7.60}$$

It is easy to eliminate y'', but not y', so the solution necessarily remains in parametric form.

Point and contact transformations are geometrical transformations that are well defined without reference to any particular ODE. Point transformations are diffeomorphisms of the plane, whereas contact transformations are diffeomorphisms of the jet space J^1 (that satisfy the contact condition). Both classes of transformations are extended to higher jet spaces by prolongation. It turns out that there are no other classes of diffeomorphisms of any finite-order jet space J^k, $k > 1$, that satisfy the prolongation formulae on the whole of J^k. However, this does not mean that every symmetry of an ODE is a contact symmetry. We are only concerned with the action of diffeomorphisms on the subset $\mathcal{S}$ of J^n defined by the ODE $y^{(n)} = \omega$. Therefore we do not need the prolongation conditions to hold on all of J^n, as long as they are satisfied on $\mathcal{S}$.

Any generator X whose characteristic Q satisfies the linearized symmetry condition (7.44) is said to generate *dynamical symmetries* (or *internal symmetries*). If Q depends upon any derivative $y^{(k)}$, $k > 1$, the dynamical symmetries are not contact symmetries.

Note that $X = \xi D_x$ generates dynamical symmetries for all ξ, because the characteristic corresponding to X is $Q = 0$ (irrespective of ξ). This is a trivial solution of the linearized symmetry condition; every solution is invariant, and so these symmetries are of no use. Two dynamical symmetry generators X_1 and X_2 are said to be equivalent if

$$X_1 - X_2 = \xi D_x \qquad \text{when} \quad y^{(n)} = \omega,$$

for some function ξ. Equivalent generators have identical characteristics. Hence, without loss of generality, we may restrict attention to dynamical symmetries whose generators have no ∂_x term. These generators, restricted to the solutions of $y^{(n)} = \omega$, are of the form

$$X = Q\partial_y + \bar{D}Q\partial_{y'} + \cdots + \bar{D}^{n-1}Q\partial_{y^{(n-1)}}. \tag{7.61}$$

Henceforth we shall assume that generators of dynamical symmetries have the form (7.61).

If $Q = Q_0$ is a solution of the linearized symmetry condition

$$\bar{D}^n Q - \omega_{y^{(n-1)}} \bar{D}^{n-1} Q - \cdots - \omega_{y'} \bar{D} Q - \omega_y Q = 0, \tag{7.62}$$

then so is $Q = \phi Q_0$, for any first integral ϕ. Given a set of n functionally independent first integrals, $\phi^1, \ldots, \phi^n$, one can construct a set of generators $X_1, \ldots, X_n$ such that

$$X_i \phi^j = \delta_i^j .$$

(Here δ_i^j is the Kronecker delta.) If $Q_1, \ldots, Q_n$ are the characteristics corresponding to these generators, then the general solution of (7.62) is

$$Q = F^i(\phi^1, \ldots, \phi^n) Q_i, \tag{7.63}$$

where $F^1, \ldots, F^n$ are arbitrary functions of the first integrals. If we use $(x, \phi^1, \ldots, \phi^n)$ as coordinates, the generators X_i reduce to

$$X_i = \partial_{\phi^i},$$

which are translations. So (up to equivalence) every dynamical symmetry generator is of the form

$$X = F^i(\phi^1, \ldots, \phi^n) \partial_{\phi^i}. \tag{7.64}$$

Thus dynamical symmetries of ODEs can be regarded as transformations of the set of first integrals.

Normally one cannot find all dynamical symmetries without first solving the ODE. But, just as for contact symmetries, it is possible to seek those dynamical symmetries that depend on $x, y, y', \ldots, y^{(n-1)}$ in some specific way. For instance, it is often fruitful to look for characteristics that are independent of $y^{(n-1)}$. If a sufficient number of dynamical symmetries can be found, the method of §7.1 enables the user to construct first integrals.

Example 7.5 Consider the fourth-order ODE

$$y'''' = \frac{y''' + y'''^2}{y''}, \tag{7.65}$$

which has a four-dimensional Lie algebra of point symmetry generators. Although this ODE can be solved by using differential invariants, it is instructive to try to use the method of §7.1. There are six linearly independent characteristics corresponding to dynamical symmetries that are independent of y''':

$$\begin{gathered} Q_1 = 1, \qquad Q_2 = y', \qquad Q_3 = x, \qquad Q_4 = 3y - xy', \\ Q_5 = y'', \qquad Q_6 = xy'' - \tfrac{1}{2}x^2. \end{gathered} \tag{7.66}$$

(These are found by equating powers of y''' in the linearized symmetry condition.) Although $\nu = 4$, there are only three first integrals that can be obtained by taking ratios of determinants W_{ijkl}. These are:

$$\begin{aligned} \phi^1 &= \frac{1 + y'''}{y''}, \\ \phi^2 &= y''' + x\phi^1 - y'(\phi^1)^2, \\ \phi^3 &= y'' - xy''' - \tfrac{1}{2}x^2\phi^1 + (xy' - y)(\phi^1)^2. \end{aligned} \tag{7.67}$$

At first sight, the relationship

$$\bar{D}W_{ijkl} = \omega_{y'''}W_{ijkl} = \frac{1 + 2y'''}{y''}W_{ijkl} \tag{7.68}$$

does not seem to provide the extra first integral. However, we have the freedom to substitute any of the known first integrals into (7.68). For instance, we may rewrite (7.68) as

$$\bar{D}W_{ijkl} = \left(\phi^1 + \frac{y'''}{y''}\right)W_{ijkl} = \bar{D}(\phi^1 x + \ln|y''|)W_{ijkl}.$$

Substituting $W_{1234} = (y''' + y'''^2)$ into the above yields the first integral

$$\phi^4 = \ln\left|\frac{y''' + y'''^2}{y''}\right| - \phi^1 x. \tag{7.69}$$

The general solution of the ODE is obtainable in closed form by eliminating y', y'', and y''':

$$y = \frac{1}{2c_1}x^2 + c_2 x + c_3 + c_4 e^{c_1 x}. \tag{7.70}$$

7.3 Integrating Factors

The method outlined in §7.1 is useful because it can produce the general solution of a given ODE, without any need for quadrature. Furthermore, it does not depend on the existence of an n-dimensional solvable subalgebra; there is no direct use of the algebraic structure. However, the method is limited by the need for at least $n+1$ symmetry generators.

Some ODEs that can be solved easily lack Lie point symmetries but have an obvious first integral, ϕ^1. If the reduced ODE $\phi^1 = c_1$ has symmetries that can be found, or an obvious first integral, then further reduction of order is possible.

Example 7.6 Consider the class of ODEs of the form

$$y'' = \frac{y'^2}{y} + f(x)yy' + f'(x)y^2. \tag{7.71}$$

These ODEs have no Lie point symmetries unless there are constants $k_1, \ldots, k_4$ (not all of which are zero) such that

$$(k_1 x + k_2)f'(x) + (k_3 x + k_4)f(x) = 0.$$

Nevertheless, every ODE (7.71) can be reduced to quadrature, as follows. Multiply (7.71) by y^{-1} and integrate once to obtain

$$\phi^1 \equiv \frac{y'}{y} - f(x)y = c_1. \tag{7.72}$$

The first-order ODE (7.72) is a Bernoulli equation for y, and is equivalent to

$$\left(\frac{1}{y}\right)' + \frac{c_1}{y} = -f(x).$$

This can be solved with an integrating factor (using the symmetries associated with linear superposition). Thus the general solution of (7.71) is

$$y = \frac{e^{c_1 x}}{c_2 - \int f(x)e^{c_1 x}\,dx}. \tag{7.73}$$

The key step in solving the above example was the multiplication of the ODE by y^{-1}. This method generalizes to arbitrary ODEs as follows. A (nonzero) function Λ is an *integrating factor* for the ODE (7.1) if

$$\left(y^{(n)} - \omega\right)\Lambda = D_x \chi \tag{7.74}$$

for some function $\chi(x, y, y', \ldots, y^{(n-1)})$; thus χ is a first integral. Indeed, every first integral ϕ satisfies

$$D_x\phi = \bar{D}\phi + \left(y^{(n)} - \omega\right)\phi_{y^{(n-1)}} = \left(y^{(n)} - \omega\right)\phi_{y^{(n-1)}}. \tag{7.75}$$

Therefore every integrating factor of (7.1) is of the form

$$\Lambda\left(x, y, y', \ldots, y^{(n-1)}\right) \equiv \phi_{y^{(n-1)}} \tag{7.76}$$

for some first integral ϕ.

Integrating factors can be determined systematically, in much the same way as symmetries are found from (7.62). The following identities are useful:

$$\bar{D}\partial_{y^{(k)}} = \partial_{y^{(k)}}\bar{D} - \partial_{y^{(k-1)}} - \omega_{y^{(k)}}\partial_{y^{(n-1)}}, \qquad k = 0, \ldots, n-1; \tag{7.77}$$

here we adopt the conventions $y = y^{(0)}$ and $\partial_{y^{(-1)}} \equiv 0$. Equation (7.76) gives

$$\phi_{y^{(n-1)}} = \Lambda. \tag{7.78}$$

We now apply the operator $\bar{D}$ to $\phi_{y^{(n-1)}}$. Using the identity (7.77) with $k = n-1$, and taking $\bar{D}\phi = 0$ into account, we obtain

$$\phi_{y^{(n-2)}} = -\left(\bar{D}\phi_{y^{(n-1)}} + \omega_{y^{(n-1)}}\phi_{y^{(n-1)}}\right). \tag{7.79}$$

Similarly, applying $\bar{D}$ to each partial derivative $\phi_{y^{(k)}}$ in turn yields

$$\phi_{y^{(k-1)}} = -\left(\bar{D}\phi_{y^{(k)}} + \omega_{y^{(k)}}\phi_{y^{(n-1)}}\right), \qquad k = 0, \ldots, n-1. \tag{7.80}$$

In particular, the $k = 0$ equation is

$$0 = -\left(\bar{D}\phi_y + \omega_y\phi_{y^{(n-1)}}\right), \tag{7.81}$$

so the sequence terminates. Furthermore, the relationship $\bar{D}\phi = 0$ gives

$$\phi_x = -y'\phi_y - y''\phi_{y'} - \cdots - y^{(n-1)}\phi_{y^{(n-2)}} - \omega\phi_{y^{(n-1)}}. \tag{7.82}$$

Thus it is possible to use (7.78) to write each partial derivative of ϕ in terms of Λ and its derivatives, and (7.81) yields

$$\begin{aligned}\bar{D}^{n}\Lambda + \bar{D}^{n-1}\left(\omega_{y^{(n-1)}}\Lambda\right) - \bar{D}^{n-2}\left(\omega_{y^{(n-2)}}\Lambda\right) + \cdots \\ + (-1)^{n-1}\omega_y\Lambda = 0.\end{aligned} \tag{7.83}$$

Note that the first two terms have the same sign, whereas subsequent terms alternate in sign. Equation (7.83) is the *adjoint* of the linearized symmetry condition (7.62), and so its solutions have been called *adjoint symmetries*. This is a rather misleading name, because they are neither symmetries nor generators of symmetries. Instead, we shall call any nonzero solution Λ of (7.83) a *cocharacteristic*. By construction, every integrating factor is a cocharacteristic. However, the converse is not true, as is shown below.

Cocharacteristics are found by the same strategy as is used to find characteristics for dynamical symmetries, namely by supposing that Λ is of a particular form. For example, we may systematically search for all solutions of (7.83) that are independent of $y^{(n-1)}$. After the solution(s) Λ^i have been found, it is easy to find out which (if any) are integrating factors. First calculate (recursively) the quantities

$$\begin{aligned} P^i_{n-1} &= \Lambda^i, \\ P^i_{k-1} &= -\bar{D}P^i_k - \omega_{y^{(k)}}\Lambda^i, \qquad k = n-1, n-2, \ldots, 1. \end{aligned} \tag{7.84}$$

From (7.78), (7.80), and (7.82), we see that Λ^i is an integrating factor if

$$\begin{aligned} P^i_k = \phi^i_{y^{(k)}}, \qquad k = 0, \ldots, n-1, \quad \text{and} \\ \omega P^i_{n-1} + \textstyle\sum_{k=0}^{n-2} y^{(k+1)} P^i_k = -\phi^i_x, \end{aligned} \tag{7.85}$$

for some function ϕ^i. The integrability conditions $\phi^i_{y^{(j)}y^{(k)}} = \phi^i_{y^{(k)}y^{(j)}}$ and $\phi^i_{y^{(k)}x} = \phi^i_{xy^{(k)}}$ amount to

$$\frac{\partial P^i_k}{\partial y^{(j)}} = \frac{\partial P^i_j}{\partial y^{(k)}}, \qquad 0 \le j < k \le n-1,$$

$$\frac{\partial P^i_j}{\partial x} = -\frac{\partial}{\partial y^{(j)}}\left(\omega P^i_{n-1} + \sum_{k=0}^{n-2} y^{(k+1)} P^i_k\right), \qquad 0 \le j \le n-1.$$

Perhaps surprisingly, these conditions are all satisfied if and only if

$$\frac{\partial P^i_{n-1}}{\partial y^{(j)}} = \frac{\partial P^i_j}{\partial y^{(n-1)}}, \qquad 0 \le j \le n-2. \tag{7.86}$$

(The derivation of this result is left as an exercise.) Thus Λ^i is an integrating factor if and only if the integrability conditions (7.86) are satisfied. Specifically, ϕ^i is obtained as a line integral from

$$\begin{aligned} \phi^i &= \int \phi^i_x\, dx + \phi^i_y\, dy + \phi^i_{y'}\, dy' + \cdots + \phi^i_{y^{(n-1)}}\, dy^{(n-1)} \\ &= \int P^i_0(dy - y'\,dx) + P^i_1(dy' - y''\,dx) + \cdots + P^i_{n-1}\left(dy^{(n-1)} - \omega\, dx\right). \end{aligned} \tag{7.87}$$

Example 7.7 The third-order ODE

$$y''' = 3yy' \tag{7.88}$$

has a two-parameter Lie group of point symmetries, with generators

$$X_1 = \partial_x, \qquad X_2 = x\partial_x - 2y\partial_y;$$

these are the only Lie contact symmetries. The method of differential invariants leads to the first-order ODE

$$\frac{dv_2}{dr_2} = \frac{2r_2(3 - 2v_2)}{2v_2 - 3r_2}, \qquad (r_2, v_2) = (y^{-3/2}y',\ y^{-2}y''),$$

but this appears to be intractable. Equation (7.83) for the integrating factors amounts to

$$\bar{D}^3\Lambda - \bar{D}(3y\Lambda) + 3y'\Lambda = 0. \tag{7.89}$$

Just as for contact symmetries, we find the solutions of the form $\Lambda = \Lambda(x, y, y')$ by equating powers of y'' in (7.89) and then solving the resulting overdetermined set of PDEs. This yields three linearly independent solutions:

$$\Lambda^1 = 1, \qquad \Lambda^2 = y, \qquad \Lambda^3 = y'^2 - y^3.$$

From (7.84), we obtain

$$P_2^1 = 1, \qquad P_1^1 = 0, \qquad P_0^1 = -3y;$$
$$P_2^2 = y, \qquad P_1^2 = -y', \qquad P_0^2 = y'' - 3y^2;$$
$$P_2^3 = y'^2 - y^3, \qquad P_1^3 = 3y^2y' - 2y'y'',$$
$$P_0^3 = 2y''^2 - 3y^2y'' - 3yy'^2 + 3y^4.$$

The integrability conditions (7.86) are satisfied for $i = 1, 2$, so Λ^1 and Λ^2 are integrating factors. However,

$$\frac{\partial P_2^3}{\partial y'} - \frac{\partial P_1^3}{\partial y''} = 4y' \neq 0,$$

so Λ^3 is not an integrating factor. Thus we have obtained two first integrals:

$$\phi^1 = \int (dy'' - 3yy'\,dx) - 3y(dy - y'\,dx) = y'' - \tfrac{3}{2}y^2; \tag{7.90}$$

$$\begin{aligned}\phi^2 &= \int y(dy'' - 3yy'\,dx) - y'(dy' - y''\,dx) + (y'' - 3y^2)(dy - y'\,dx)\\ &= yy'' - \tfrac{1}{2}y'^2 - y^3.\end{aligned} \tag{7.91}$$

We do not yet have enough first integrals to describe the general solution of the ODE, but combining the equations $\phi^i = c_i,\ i = 1, 2$ leads to the separable first-order ODE

$$y'^2 = y^3 + 2c_1 y - 2c_2.$$

Hence the general solution of (7.88) is

$$x = c_3 \pm \int \frac{dy}{\sqrt{y^3 + 2c_1 y - 2c_2}}. \tag{7.92}$$

Another way to generate first integrals is to combine cocharacteristics and symmetries. For example, suppose that Λ is a cocharacteristic for the second-order ODE

$$y'' = \omega(x, y, y').$$

As usual, let

$$P_1 = \Lambda, \qquad P_0 = -\left(\bar{D}P_1 + \omega_{y'}\Lambda\right),$$

and note that $\bar{D}P_0 + \omega_y \Lambda = 0$. Now suppose that Q is the characteristic for any (nontrivial) symmetry of the ODE, and let

$$\Phi = P_0 Q + P_1 \bar{D} Q.$$

Then Φ is either a first integral or a constant, because

$$\begin{aligned} \bar{D}\Phi &= (\bar{D}P_0)Q + (P_0 + \bar{D}P_1)\bar{D}Q + P_1 \bar{D}^2 Q \\ &= (\bar{D}P_0 + \omega_y P_1)Q + \left(P_0 + \bar{D}P_1 + \omega_{y'} P_1\right)\bar{D}Q \\ &= 0. \end{aligned}$$

A similar result holds for every ODE $y^{(n)} = \omega$. Given any cocharacteristic Λ and any characteristic Q, the function

$$\Phi = \sum_{k=0}^{n-1} P_k \bar{D}^k Q \tag{7.93}$$

is either a first integral or a constant. This result does not depend upon the integrability conditions (7.86) being satisfied. If several symmetries are known, then each characteristic can be substituted into (7.93). Thus a single cocharacteristic may yield several functionally independent first integrals.

7.4 Systems of ODEs

The above ideas may be generalized to systems of ODEs. For simplicity, we shall concentrate on systems of n first-order ODEs

$$y_k' = \omega_k(x, y_1, \ldots, y_n), \qquad k = 1, \ldots, n. \tag{7.94}$$

There is no loss of generality in doing this, for every system of ODEs can be rewritten as an equivalent first-order system. The main results on symmetries and integrating factors are derived by the same methods as we have used for a single ODE. Therefore we state these results without detailed justification; the reader should be able to derive them.

Lie point transformations of the variables $x, y_1, \ldots, y_n$ are generated by infinitesimal generators of the form

$$X = \xi(x, y_1, \ldots, y_n)\partial_x + \eta_k(x, y_1, \ldots, y_n)\partial_{y_k}. \tag{7.95}$$

(Here the usual summation convention applies.) The *characteristic* of the one-parameter Lie group generated by X is $\mathbf{Q} = (Q_1, \ldots, Q_n)$, where

$$Q_k = \eta_k - y_k'\xi, \qquad k = 1, \ldots, n.$$

To find Lie point symmetries of the system (7.94), we prolong the infinitesimal generator X to first derivatives as follows:

$$X^{(1)} = \xi\partial_x + \eta_k\partial_{y_k} + \eta_k^{(1)}\partial_{y_k'},$$

where

$$\eta_k^{(1)} = D_x\eta_k - y_k' D_x\xi = D_x Q_k + y_k''\xi, \qquad k = 1, \ldots, n. \tag{7.96}$$

Then the linearized symmetry condition is

$$X^{(1)}(y_k' - \omega_k) = 0, \qquad k = 1, \ldots, n, \qquad \text{when (7.94) holds.} \tag{7.97}$$

This condition may also be written in terms of the *reduced characteristic*, $\bar{\mathbf{Q}} = (\bar{Q}_1, \ldots, \bar{Q}_n)$, whose components are

$$\bar{Q}_k(x, y_1, \ldots, y_k) \equiv \eta_k - \omega_k\xi, \qquad k = 1, \ldots, n.$$

With a little work, it can be shown that the linearized symmetry condition amounts to

$$\bar{D}\bar{Q}_k - \frac{\partial\omega_k}{\partial y_j}\bar{Q}_j = 0, \qquad k = 1, \ldots, n, \tag{7.98}$$

where

$$\bar{D} = \partial_x + \omega_i \partial_{y_i}.$$

A solution is invariant under a one-parameter Lie group of symmetries if the reduced characteristic is zero on the solution. The symmetry group is trivial if every solution is invariant, that is, if $\bar{Q}_k \equiv 0,\ k = 1, \ldots, n$. Therefore a one-parameter Lie group of symmetries is trivial if and only if its infinitesimal generator is of the form

$$X = \xi \bar{D},$$

for some function ξ. Any symmetry generator (7.95) may be split into its trivial and nontrivial components as follows:

$$X = \xi \bar{D} + \bar{Q}_k \partial_{y_k}.$$

Therefore the difference between any two generators with the same $\bar{\mathbf{Q}}$ is the generator of a trivial symmetry. For now, let us exclude trivial symmetries by setting $\xi = 0$ and considering only those generators that are of the form

$$X = \bar{Q}_k \partial_{y_k}. \tag{7.99}$$

Any single ODE $y^{(n)} = \omega$ may be represented (in many ways) by an equivalent system of n first-order ODEs. (Here "equivalent" means that the general solution is the same.) We now show that there is a one-to-one correspondence between the nontrivial dynamical symmetries of $y^{(n)} = \omega$ and the Lie point symmetries (7.99) of any equivalent first-order system.

A nonconstant function $\phi(x, y_1, \ldots, y_n)$ is a first integral of (7.94) if

$$\bar{D}\phi = 0. \tag{7.100}$$

The general solution of (7.94) depends on n arbitrary constants, and thus there are n functionally independent first integrals, $\phi^1, \ldots, \phi^n$. The general solution is

$$\phi^k = c_k, \qquad k = 1, \ldots, n,$$

so the system of ODEs (7.94) is equivalent to

$$\frac{d\phi^k}{dx} = 0, \qquad k = 1, \ldots, n. \tag{7.101}$$

Assuming that the change of dependent variables from y_k to ϕ^k is a diffeomorphism, the Lie point symmetries of (7.94) correspond to Lie point symmetries of the equivalent system (7.101). Under this change of variables, the generator (7.99) is (by the chain rule)

$$X = F^i(x, \phi^1, \dots, \phi^n)\partial_{\phi^i}, \qquad \text{where} \quad F^i = \bar{Q}_k \phi^i_{y_k}. \tag{7.102}$$

The chain rule also yields

$$\bar{D} = \partial_x.$$

Thus the linearized symmetry condition for (7.101) is

$$\bar{D}F^i = F^i_x = 0, \qquad i = 1, \dots, n.$$

Consequently, every Lie point symmetry generator (7.102) is of the form

$$X = F^i(\phi^1, \dots, \phi^n)\partial_{\phi^i}, \tag{7.103}$$

for some (arbitrary) functions $F^1, \dots, F^n$.

If (7.94) is equivalent to a single ODE $y^{(n)} = \omega$, that ODE has the same general solution, and thus the same set of first integrals. Comparing (7.103) with (7.64), we see that the Lie point symmetries of the system (7.94) are the same as the dynamical symmetries of the equivalent nth order ODE.

There are infinitely many Lie point symmetries of any given first-order system of ODEs, but generally one cannot find them without first solving the system. Some sort of ansatz is needed; commonly, one seeks reduced characteristics that are independent of (or linear in) one variable. Alternatively, we can substitute a particular ansatz for $(\xi, \eta_1, \dots, \eta_n)$ into (7.97). For some of the most common point transformations (e.g., translations, scalings, and rotations), ξ and η_k are linear in each variable.

Once a nontrivial symmetry generator has been found, it is possible (in principle) to reduce the problem of solving (7.94) to that of solving $n-1$ first-order ODEs and then carrying out a final quadrature. To do this, introduce canonical coordinates $(s, r_1, \dots, r_n)$ such that

$$Xr_k = 0, \qquad k = 1, \dots, n, \qquad Xs = 1.$$

Then $X = \partial_s$, and so the system of ODEs is equivalent to

$$\frac{dr_k}{ds} = \Omega_k(r_1, \dots, r_n), \qquad k = 1, \dots, n \tag{7.104}$$

for some functions Ω_k. If these functions are all zero, then (7.104) can be solved immediately. Otherwise, assume (without loss of generality) that $\Omega_1 \neq 0$. Then (7.104) yields $n-1$ ODEs involving only the differential invariants r_k:

$$\frac{dr_k}{dr_1} = \frac{\Omega_k}{\Omega_1}, \qquad k = 2, \ldots, n. \tag{7.105}$$

If the solution of (7.105) gives $r_2, \ldots, r_n$ in terms of r_1, the remaining ODE in (7.104) can be solved by quadrature:

$$s = \int \frac{dr_1}{\Omega_1\big(r_1,\, r_2(r_1),\, \ldots,\, r_n(r_1)\big)} + c_n. \tag{7.106}$$

If the solution of (7.105) is parametric, s is obtained by rewriting the integral in (7.106) in terms of the parameter.

The above method requires the user to find differential invariants of the given one-parameter group. In practice this is not always possible, because the characteristic equations

$$\frac{dx}{\xi} = \frac{dy_1}{\eta_1} = \cdots = \frac{dy_n}{\eta_n} = ds \tag{7.107}$$

may be hard to solve. The method is most successful for ODEs that have symmetry generators of a fairly simple form.

The Lie point symmetry generators of a given first-order system of ODEs form an infinite-dimensional Lie algebra. As the commutator of any two generators is a generator, one may be able to construct new generators from known ones. Usually this process terminates quickly, and one is left with a finite-dimensional subalgebra, $\mathcal{L}$. If $\mathcal{L}$ has an solvable subalgebra of dimension $R \leq n$, the system of ODEs (7.94) may be reduced to $n - R$ ODEs in the differential invariants, together with R quadratures. The procedure is similar to that for a single nth order ODE.

Systems of ODEs can also be solved by using integrating factors to construct first integrals. If $\phi(x, y_1, \ldots, y_n)$ is a first integral of the system (7.94) then $\bar{D}\phi = 0$, and hence

$$D_x\phi = (y_k' - \omega_k)\phi_{y_k}. \tag{7.108}$$

We call $\mathbf{\Lambda} = (\Lambda^1, \ldots, \Lambda^n)$ an integrating factor of the system (7.94) if

$$(y_k' - \omega_k)\Lambda^k = D_x\phi$$

for some nonconstant function $\phi(x, y_1, \ldots, y_n)$. Therefore the components of every integrating factor satisfy

$$\phi_{y_k} = \Lambda^k, \qquad k = 1, \ldots, n. \tag{7.109}$$

Applying $\bar{D}$ to (7.109) and taking $\bar{D}\phi = 0$ into account, we obtain

$$\bar{D}\Lambda^k + \frac{\partial \omega_i}{\partial y_k}\Lambda^i = 0, \qquad k = 1, \ldots, n. \tag{7.110}$$

Each nonzero solution Λ of (7.110) is called a *cocharacteristic*. A cocharacteristic is an integrating factor if and only it satifies the integrability conditions

$$\frac{\partial \Lambda^k}{\partial y_j} = \frac{\partial \Lambda^j}{\partial y_k}, \qquad 1 \leq j < k \leq n. \tag{7.111}$$

Given any integrating factor, we use (7.108) to reconstruct the corresponding first integral as follows:

$$\phi = \int \Lambda^k (dy_k - \omega_k \, dx). \tag{7.112}$$

Another way to construct first integrals is to combine characteristics and cocharacteristics, just as we did for single ODEs. From (7.98) and (7.110), we find that

$$\bar{D}\big(\bar{Q}_k \Lambda^k\big) = 0. \tag{7.113}$$

Hence $\Phi = \bar{Q}_k \Lambda^k$ is either a constant or a first integral.

Example 7.8 Consider the system of ODEs

$$\begin{aligned} y_1' &= \frac{xy_1 + y_2^2}{y_1 y_2 - x^2}, \\ y_2' &= \frac{xy_2 + y_1^2}{y_1 y_2 - x^2}. \end{aligned} \tag{7.114}$$

This system has the scaling symmetries generated by

$$X = x\partial_x + y_1\partial_{y_1} + y_2\partial_{y_2};$$

this is the only infinitesimal generator whose coefficients are linear in y_1 and y_2. The characteristic equations (7.107) are easily solved to obtain the canonical coordinates

$$(s, r_1, r_2) = (\ln|x|, \; y_1/x, \; y_2/x).$$

Therefore the system (7.114) is equivalent to

$$\frac{dr_1}{ds} = \frac{2r_1 + r_2^2 - r_1^2 r_2}{r_1 r_2 - 1},$$
$$\frac{dr_2}{ds} = \frac{2r_2 + r_1^2 - r_1 r_2^2}{r_1 r_2 - 1}.$$

As promised, this yields one ODE that is independent of s:

$$\frac{dr_2}{dr_1} = \frac{2r_2 + r_1^2 - r_1 r_2^2}{2r_1 + r_2^2 - r_1^2 r_2}.$$

However this ODE has no obvious symmetries; it does not seem easy to solve.

The system (7.114) has two linearly independent cocharacteristics whose components are linear functions of y_1 and y_2; these are

$$\mathbf{\Lambda}_1 = (y_1, -x), \qquad \mathbf{\Lambda}_2 = (-x, y_2).$$

Each of these satisfy the integrability conditions (7.111), and (7.112) enables us to construct the first integrals

$$\begin{aligned} \phi^1 &= \tfrac{1}{2}y_1^2 - xy_2, \\ \phi^2 &= \tfrac{1}{2}y_2^2 - xy_1. \end{aligned} \tag{7.115}$$

Therefore the general solution of the system (7.114) is

$$\begin{aligned} y_1^4 - 4c_1 y_1^2 - 8x^3 y_1 - 8c_2 x^2 + 4c_1^2 = 0, \\ y_2^4 - 4c_2 y_2^2 - 8x^3 y_2 - 8c_1 x^2 + 4c_2^2 = 0. \end{aligned} \tag{7.116}$$

Further Reading

Dynamical symmetries are described in greater detail by Stephani (1989), who includes several examples from classical physics.

The integrating factor method of §7.3 is a modified version of the technique described in Anco and Bluman (1998). These authors have also addressed the related problem of constructing conservation laws of PDEs.

Exercises

7.1 Find three functionally independent first integrals of the ODE

$$y''' = \frac{6y'y''}{y} - \frac{6y'^3}{y^2},$$

using the symmetries generated by

$$X_1 = \partial_x, \qquad X_2 = y^2\partial_y, \qquad X_3 = xy^2\partial_y, \qquad X_4 = x\partial_x - 2y\partial_y.$$

Use the first integrals to determine the general solution of the ODE.

7.2 Find all characteristics of contact symmetries of the ODE

$$y''' = 3y''^2/(2y') + 2y'^3$$

and use them to construct the first integrals.

7.3 Derive all characteristics of contact symmetries of the ODE

$$y''' = x(x-1)y''^3 - 2xy''^2 + y''$$

and use them to construct the first integrals (7.60).

7.4 Given an infinitesimal generator of Lie contact symmetries, one can try to reconstruct the finite contact symmetries by solving

$$\frac{d\hat{x}}{d\varepsilon} = \xi(\hat{x}, \hat{y}, \hat{y}'), \qquad \frac{d\hat{y}}{d\varepsilon} = \eta(\hat{x}, \hat{y}, \hat{y}'), \qquad \frac{d\hat{y}'}{d\varepsilon} = \eta^{(1)}(\hat{x}, \hat{y}, \hat{y}'),$$

subject to the initial conditions

$$(\hat{x}, \hat{y}, \hat{y}') = (x, y, y') \qquad \text{when} \quad \varepsilon = 0.$$

Find the finite contact symmetries corresponding to each characteristic in (7.56).

7.5 Use the dynamical symmetries (7.66) to derive the first integrals (7.67).

7.6 Show that the integrability conditions (7.86) are sufficient to ensure that all integrability conditions are satisfied. [Hint: Let

$$J_j^k = \frac{\partial P_k^i}{\partial y^{(j)}} - \frac{\partial P_j^i}{\partial y^{(k)}}$$

and apply $\bar{D}$ to the integrability condition $J_j^k = 0$.]

7.7 Find all characteristics and cocharacteristics depending on (x, y, y') for the ODE

$$y''' = (2y' - 1)y'' + y'^2.$$

Hence reduce the ODE to the first-order (Riccati) equation

$$v' - v^2 = c_1 e^x.$$

(The solutions of this equation can be written in terms of Bessel functions, as follows. First substitute $v = -z'/z$, to linearize the Riccati equation; then introduce a new independent variable $t = e^{-x/2}$, to obtain Bessel's equation.) Now determine the general solution of the original third-order ODE.

7.8 Show that if Φ is defined by (7.93) then $\bar{D}\Phi = 0$.

7.9 Use the first integrals (7.41), (7.42) to construct the generators $X_i = \partial_{\phi^i}$. [You should find that X_3 is a dynamical symmetry.] Write the characteristic corresponding to each point symmetry generator (7.40) in the form (7.63). Use your results to explain why the first integral (7.42) could not be found by taking ratios of the determinants W_{ijk}.

8

How to Obtain Lie Point Symmetries of PDEs

Yet what are all such gaities to me
Whose thoughts are full of indices . . . ?

(Lewis Carroll: Four Riddles)

8.1 Scalar PDEs with Two Dependent Variables

Point symmetries of PDEs are defined in much the same way as those of ODEs. For simplicity, let us start by considering PDEs with one dependent variable, u, and two independent variables, x and t. A point transformation is a diffeomorphism

$$\Gamma : (x, t, u) \mapsto \big(\hat{x}(x, t, u),\ \hat{t}(x, t, u),\ \hat{u}(x, t, u)\big). \tag{8.1}$$

This transformation maps the surface $u = u(x, t)$ to the following surface (which is parametrized by x and t):

$$\begin{aligned} \hat{x} &= \hat{x}\big(x, t, u(x, t)\big), \\ \hat{t} &= \hat{t}\big(x, t, u(x, t)\big), \\ \hat{u} &= \hat{u}\big(x, t, u(x, t)\big). \end{aligned} \tag{8.2}$$

To calculate the prolongation of a given transformation, we need to differentiate (8.2) with respect to each of the parameters x and t. To do this, we introduce the following *total derivatives*:

$$\begin{aligned} D_x &= \partial_x + u_x \partial_u + u_{xx} \partial_{u_x} + u_{xt} \partial_{u_t} + \cdots, \\ D_t &= \partial_t + u_t \partial_u + u_{xt} \partial_{u_x} + u_{tt} \partial_{u_t} + \cdots. \end{aligned} \tag{8.3}$$

(Total derivatives treat the dependent variable u and its derivatives as functions of the independent variables.)

The first two equations of (8.2) may be inverted (locally) to give x and t in terms of $\hat{x}$ and $\hat{t}$, provided that the Jacobian is nonzero, that is,

$$\mathcal{J} \equiv \begin{vmatrix} D_x\hat{x} & D_x\hat{t} \\ D_t\hat{x} & D_t\hat{t} \end{vmatrix} \neq 0 \quad \text{when} \quad u = u(x,t). \tag{8.4}$$

If (8.4) is satisfied, then the last equation of (8.2) can be rewritten as

$$\hat{u} = \hat{u}(\hat{x},\hat{t}). \tag{8.5}$$

Applying the chain rule to (8.5), we obtain

$$\begin{bmatrix} D_x\hat{u} \\ D_t\hat{u} \end{bmatrix} = \begin{bmatrix} D_x\hat{x} & D_x\hat{t} \\ D_t\hat{x} & D_t\hat{t} \end{bmatrix} \begin{bmatrix} \hat{u}_{\hat{x}} \\ \hat{u}_{\hat{t}} \end{bmatrix},$$

and therefore (by Cramer's rule)

$$\hat{u}_{\hat{x}} = \frac{1}{\mathcal{J}} \begin{vmatrix} D_x\hat{u} & D_x\hat{t} \\ D_t\hat{u} & D_t\hat{t} \end{vmatrix}, \qquad \hat{u}_{\hat{t}} = \frac{1}{\mathcal{J}} \begin{vmatrix} D_x\hat{x} & D_x\hat{u} \\ D_t\hat{x} & D_t\hat{u} \end{vmatrix}. \tag{8.6}$$

Higher-order prolongations are obtained recursively by repeating the above argument. If $\hat{u}_J$ is any derivative of $\hat{u}$ with respect to $\hat{x}$ and $\hat{t}$, then

$$\begin{aligned} \hat{u}_{J\hat{x}} &\equiv \frac{\partial \hat{u}_J}{\partial \hat{x}} = \frac{1}{\mathcal{J}} \begin{vmatrix} D_x\hat{u}_J & D_x\hat{t} \\ D_t\hat{u}_J & D_t\hat{t} \end{vmatrix}, \\ \hat{u}_{J\hat{t}} &\equiv \frac{\partial \hat{u}_J}{\partial \hat{t}} = \frac{1}{\mathcal{J}} \begin{vmatrix} D_x\hat{x} & D_x\hat{u}_J \\ D_t\hat{x} & D_t\hat{u}_J \end{vmatrix}. \end{aligned} \tag{8.7}$$

For example, the transformation is prolonged to second derivatives as follows:

$$\begin{aligned} \hat{u}_{\hat{x}\hat{x}} &= \frac{1}{\mathcal{J}} \begin{vmatrix} D_x\hat{u}_{\hat{x}} & D_x\hat{t} \\ D_t\hat{u}_{\hat{x}} & D_t\hat{t} \end{vmatrix}, \qquad \hat{u}_{\hat{t}\hat{t}} = \frac{1}{\mathcal{J}} \begin{vmatrix} D_x\hat{x} & D_x\hat{u}_{\hat{t}} \\ D_t\hat{x} & D_t\hat{u}_{\hat{t}} \end{vmatrix}, \\ \hat{u}_{\hat{x}\hat{t}} &= \frac{1}{\mathcal{J}} \begin{vmatrix} D_x\hat{u}_{\hat{t}} & D_x\hat{t} \\ D_t\hat{u}_{\hat{t}} & D_t\hat{t} \end{vmatrix} = \frac{1}{\mathcal{J}} \begin{vmatrix} D_x\hat{x} & D_x\hat{u}_{\hat{x}} \\ D_t\hat{x} & D_t\hat{u}_{\hat{x}} \end{vmatrix}. \end{aligned} \tag{8.8}$$

We are now in a position to define point symmetries of an nth order PDE:

$$\Delta(x,t,u,u_x,u_t,\ldots) = 0. \tag{8.9}$$

For simplicity, we shall restrict attention to PDEs of the form

$$\Delta = u_\sigma - \omega(x,t,u,u_x,u_t,\ldots) = 0, \tag{8.10}$$

where u_σ is one of the nth order derivatives of u and ω is independent of u_σ. (More generally, u_σ could be of order $k < n$ provided that ω does not depend upon u_σ or any derivatives of u_σ.)

The point transformation Γ is a point symmetry of (8.9) if

$$\Delta(\hat{x}, \hat{t}, \hat{u}, \hat{u}_{\hat{x}}, \hat{u}_{\hat{t}}, \ldots) = 0 \qquad \text{when (8.9) holds.} \tag{8.11}$$

Typically, the symmetry condition (8.11) is extremely complicated, so we shall not try to solve it directly. Nevertheless, it is quite easy to check whether or not a given point transformation is a symmetry of a particular PDE.

Example 8.1 Here we use the symmetry condition to show that

$$(\hat{x}, \hat{t}, \hat{u}) = \left(\frac{x}{2t}, \frac{-1}{4t}, 2(ut - x)\right) \tag{8.12}$$

is a point symmetry of *Burgers' equation*

$$u_t + uu_x = u_{xx}. \tag{8.13}$$

The Jacobian of the point transformation (8.12) is

$$\mathcal{J} = \begin{vmatrix} \frac{1}{2t} & 0 \\ \frac{-x}{2t^2} & \frac{1}{4t^2} \end{vmatrix} = \frac{1}{8t^3},$$

and therefore

$$\hat{u}_{\hat{x}} = 8t^3 \begin{vmatrix} 2(tu_x - 1) & 0 \\ 2(tu_t + u) & \frac{1}{4t^2} \end{vmatrix} = 4t(tu_x - 1),$$

$$\hat{u}_{\hat{t}} = 8t^3 \begin{vmatrix} \frac{1}{2t} & 2(tu_x - 1) \\ \frac{-x}{2t^2} & 2(tu_t + u) \end{vmatrix} = 8t(t^2u_t + xtu_x + tu - x),$$

$$\hat{u}_{\hat{x}\hat{x}} = 8t^3 \begin{vmatrix} 4t^2u_{xx} & 0 \\ 4(t^2u_{xt} + 2tu_x - 1) & \frac{1}{4t^2} \end{vmatrix} = 8t^3u_{xx}.$$

Note that

$$\hat{u}_{\hat{t}} + \hat{u}\hat{u}_{\hat{x}} = 8t^3(u_t + uu_x),$$

and hence the point transformation satisfies the symmetry condition

$$\hat{u}_{\hat{t}} + \hat{u}\hat{u}_{\hat{x}} = \hat{u}_{\hat{x}\hat{x}} \qquad \text{when} \quad u_t + uu_x = u_{xx}.$$

Generally speaking, we do not know *a priori* what form the point symmetries of a given PDE will take. However, it is usually possible to carry out a systematic search for one-parameter Lie groups of point symmetries. The technique is essentially the same as for ODEs. We seek point symmetries of the form

$$\begin{aligned}\hat{x} &= x + \varepsilon\xi(x,t,u) + O(\varepsilon^2),\\ \hat{t} &= t + \varepsilon\tau(x,t,u) + O(\varepsilon^2),\\ \hat{u} &= u + \varepsilon\eta(x,t,u) + O(\varepsilon^2).\end{aligned} \tag{8.14}$$

Just as for Lie point transformations of the plane, each one-parameter (local) Lie group of point transformations is obtained by exponentiating its infinitesimal generator, which is

$$X = \xi\partial_x + \tau\partial_t + \eta\partial_u. \tag{8.15}$$

Equivalently, we can obtain $(\hat{x}, \hat{t}, \hat{u})$ by solving

$$\frac{d\hat{x}}{d\varepsilon} = \xi(\hat{x},\hat{t},\hat{u}), \qquad \frac{d\hat{t}}{d\varepsilon} = \tau(\hat{x},\hat{t},\hat{u}), \qquad \frac{d\hat{u}}{d\varepsilon} = \eta(\hat{x},\hat{t},\hat{u}),$$

subject to the initial conditions

$$(\hat{x},\hat{t},\hat{u})\big|_{\varepsilon=0} = (x,t,u).$$

A surface $u = u(x,t)$ is mapped to itself by the group of transformations generated by X if

$$X\big(u - u(x,t)\big) = 0 \qquad \text{when} \quad u = u(x,t). \tag{8.16}$$

This condition can be expressed neatly by using the *characteristic* of the group, which is

$$Q = \eta - \xi u_x - \tau u_t. \tag{8.17}$$

From (8.16), the surface $u = u(x,t)$ is invariant provided that

$$Q = 0 \qquad \text{when} \quad u = u(x,t). \tag{8.18}$$

Equation (8.18) is called the *invariant surface condition*; it is central to some of the main techniques for finding exact solutions of PDEs.

The prolongation of the point transformation (8.14) to first derivatives is

$$\begin{aligned}\hat{u}_{\hat{x}} &= u_x + \varepsilon\eta^x(x,t,u,u_x,u_t) + O(\varepsilon^2),\\ \hat{u}_{\hat{t}} &= u_t + \varepsilon\eta^t(x,t,u,u_x,u_t) + O(\varepsilon^2),\end{aligned} \tag{8.19}$$

where, from (8.6),

$$\begin{aligned}\eta^x(x,t,u,u_x,u_t) &= D_x\eta - u_x D_x\xi - u_t D_x\tau,\\ \eta^t(x,t,u,u_x,u_t) &= D_t\eta - u_x D_t\xi - u_t D_t\tau.\end{aligned} \tag{8.20}$$

The transformation is prolonged to higher-order derivatives recursively, using (8.7). Suppose that

$$\hat{u}_{\mathrm{J}} = u_{\mathrm{J}} + \varepsilon\eta^{\mathrm{J}} + O(\varepsilon^2), \tag{8.21}$$

where

$$u_{\mathrm{J}} \equiv \frac{\partial^{j_1+j_2}u}{\partial x^{j_1}\partial t^{j_2}}, \qquad \hat{u}_{\mathrm{J}} \equiv \frac{\partial^{j_1+j_2}\hat{u}}{\partial\hat{x}^{j_1}\partial\hat{t}^{j_2}}$$

for some numbers j_1 and j_2. Then (8.7) yields

$$\begin{aligned}\hat{u}_{\mathrm{J}\hat{x}} &= u_{\mathrm{J}x} + \varepsilon\eta^{\mathrm{J}x} + O(\varepsilon^2),\\ \hat{u}_{\mathrm{J}\hat{t}} &= u_{\mathrm{J}t} + \varepsilon\eta^{\mathrm{J}t} + O(\varepsilon^2),\end{aligned} \tag{8.22}$$

where

$$\begin{aligned}\eta^{\mathrm{J}x} &= D_x\eta^{\mathrm{J}} - u_{\mathrm{J}x}D_x\xi - u_{\mathrm{J}t}D_x\tau,\\ \eta^{\mathrm{J}t} &= D_t\eta^{\mathrm{J}} - u_{\mathrm{J}x}D_t\xi - u_{\mathrm{J}t}D_t\tau.\end{aligned} \tag{8.23}$$

Alternatively, we can express the functions η^{J} in terms of the characteristic, for example,

$$\begin{aligned}\eta^x &= D_xQ + \xi u_{xx} + \tau u_{xt},\\ \eta^t &= D_tQ + \xi u_{xt} + \tau u_{tt}.\end{aligned} \tag{8.24}$$

The higher-order terms are obtained by induction on j_1 and j_2:

$$\eta^{\mathrm{J}} = D_{\mathrm{J}}Q + \xi D_{\mathrm{J}}u_x + \tau D_{\mathrm{J}}u_t, \tag{8.25}$$

where

$$D_{\mathrm{J}} \equiv D_x^{j_1}D_t^{j_2}. \tag{8.26}$$

The infinitesimal generator is prolonged to derivatives by adding all terms of the form $\eta^{\mathrm{J}}\partial_{u_{\mathrm{J}}}$ up to the desired order. For example,

$$X^{(1)} = \xi\partial_x + \tau\partial_t + \eta\partial_u + \eta^x\partial_{u_x} + \eta^t\partial_{u_t} = X + \eta^x\partial_{u_x} + \eta^t\partial_{u_t}, \tag{8.27}$$

$$X^{(2)} = X^{(1)} + \eta^{xx}\partial_{u_{xx}} + \eta^{xt}\partial_{u_{xt}} + \eta^{tt}\partial_{u_{tt}}, \tag{8.28}$$

From now on, we adopt the convention that the generator is prolonged as many times as is needed to describe the group's action on all variables. (We shall not usually refer explicitly to the order of prolongation.) To find the Lie point symmetries, we need explicit expressions for (8.23). Here are some:

$$\eta^x = \eta_x + (\eta_u - \xi_x)u_x - \tau_x u_t - \xi_u u_x^2 - \tau_u u_x u_t, \tag{8.29}$$

$$\eta^t = \eta_t - \xi_t u_x + (\eta_u - \tau_t)u_t - \xi_u u_x u_t - \tau_u u_t^2, \tag{8.30}$$

$$\begin{aligned}\eta^{xx} &= \eta_{xx} + (2\eta_{xu} - \xi_{xx})u_x - \tau_{xx}u_t + (\eta_{uu} - 2\xi_{xu})u_x^2 \\ &\quad - 2\tau_{xu}u_x u_t - \xi_{uu}u_x^3 - \tau_{uu}u_x^2 u_t + (\eta_u - 2\xi_x)u_{xx} \\ &\quad - 2\tau_x u_{xt} - 3\xi_u u_x u_{xx} - \tau_u u_t u_{xx} - 2\tau_u u_x u_{xt},\end{aligned} \tag{8.31}$$

$$\begin{aligned}\eta^{xt} &= \eta_{xt} + (\eta_{tu} - \xi_{xt})u_x + (\eta_{xu} - \tau_{xt})u_t - \xi_{tu}u_x^2 \\ &\quad + (\eta_{uu} - \xi_{xu} - \tau_{tu})u_x u_t - \tau_{xu}u_t^2 - \xi_{uu}u_x^2 u_t - \tau_{uu}u_x u_t^2 \\ &\quad - \xi_t u_{xx} - \xi_u u_t u_{xx} + (\eta_u - \xi_x - \tau_t)u_{xt} - 2\xi_u u_x u_{xt} \\ &\quad - 2\tau_u u_t u_{xt} - \tau_x u_{tt} - \tau_u u_x u_{tt},\end{aligned} \tag{8.32}$$

$$\begin{aligned}\eta^{tt} &= \eta_{tt} - \xi_{tt}u_x + (2\eta_{tu} - \tau_{tt})u_t - 2\xi_{tu}u_x u_t \\ &\quad + (\eta_{uu} - 2\tau_{tu})u_t^2 - \xi_{uu}u_x u_t^2 - \tau_{uu}u_t^3 - 2\xi_t u_{xt} \\ &\quad - 2\xi_u u_t u_{xt} + (\eta_u - 2\tau_t)u_{tt} - \xi_u u_x u_{tt} - 3\tau_u u_t u_{tt}.\end{aligned} \tag{8.33}$$

Lie point symmetries are obtained by differentiating the symmetry condition (8.11) with respect to ε at $\varepsilon = 0$. We obtain the linearized symmetry condition

$$X\Delta = 0 \qquad \text{when} \quad \Delta = 0. \tag{8.34}$$

The restriction (8.10) enables us to eliminate u_σ from (8.34); then we split the remaining terms (according to their dependence on derivatives of u) to obtain a linear system of *determining equations* for ξ, τ, and η. The vector space $\mathcal{L}$ of all Lie point symmetry generators of a given PDE is a Lie algebra, although it may not be finite dimensional.

Example 8.2 As a simple illustration of the technique, consider the PDE

$$u_t = u_x^2. \tag{8.35}$$

The linearized symmetry condition is

$$\eta^t = 2u_x\eta^x \qquad \text{when (8.35) holds.} \tag{8.36}$$

Writing this out explicitly and using (8.35) to eliminate u_t, we obtain

$$\eta_t - \xi_t u_x + (\eta_u - \tau_t)u_x^2 - \xi_u u_x^3 - \tau_u u_x^4$$
$$= 2u_x\left(\eta_x + (\eta_u - \xi_x)u_x - (\xi_u + \tau_x)u_x^2 - \tau_u u_x^3\right)$$

After equating the terms that are multiplied by each power of u_x, we are left with the system of determining equations

$$\tau_u = 0, \tag{8.37}$$

$$\xi_u + 2\tau_x = 0, \tag{8.38}$$

$$\eta_u + \tau_t - 2\xi_x = 0, \tag{8.39}$$

$$\xi_t + 2\eta_x = 0, \tag{8.40}$$

$$\eta_t = 0. \tag{8.41}$$

(These are ordered with the u_x^4 terms first, then the u_x^3 terms, etc.) We begin by solving (8.37) to obtain

$$\tau = A(x, t),$$

where A is an arbitrary function (at present). Therefore the general solution of (8.38) is

$$\xi = -2A_x u + B(x, t),$$

and (8.39) yields

$$\eta = -2A_{xx}u^2 + (2B_x - A_t)u + C(x, t),$$

for some functions B and C. Substituting these results into (8.40) and (8.41), we obtain

$$-4A_{xxx}u^2 + 4(B_{xx} - A_{xt})u + B_t + 2C_x = 0, \tag{8.42}$$

$$-2A_{xxt}u^2 + (2B_{xt} - A_{tt})u + C_t = 0. \tag{8.43}$$

The functions A, B, and C are independent of u, so (8.42) and (8.43) can be decomposed by equating powers of u, as follows:

$$C_t = 0, \tag{8.44}$$

$$B_t + 2C_x = 0, \tag{8.45}$$

$$2B_{xt} - A_{tt} = 0, \tag{8.46}$$

$$B_{xx} - A_{xt} = 0, \tag{8.47}$$

$$A_{xxt} = 0, \tag{8.48}$$

$$A_{xxx} = 0, \tag{8.49}$$

Using each of (8.44), (8.45), and (8.46) in turn, we obtain

$$\begin{aligned} C = \alpha(x), \qquad B &= -2\alpha'(x)t + \beta(x), \\ A &= -2\alpha''(x)t^2 + \gamma(x)t + \delta(x). \end{aligned} \tag{8.50}$$

Here α, β, γ, δ are functions of x that are determined by substituting (8.50) into (8.47), (8.48), and (8.49), then equating powers of t, and solving the resulting ODEs. We eventually arrive at the general solution

$$\begin{aligned} \xi &= -4c_1tx - 2c_2t + c_4\left(\tfrac{1}{2}x^2 - 2tu\right) + c_6x + c_7 - 4c_8xu - 2c_9u, \\ \tau &= -4c_1t^2 + c_4xt + c_5t + c_8x^2 + c_9x + c_{10}, \\ \eta &= c_1x^2 + c_2x + c_3 + c_4xu - c_5u + 2c_6u - 4c_8u^2. \end{aligned} \tag{8.51}$$

There are ten arbitrary constants, signifying that the Lie algebra is ten dimensional.

As we have seen, Lie point symmetries of PDEs and ODEs are found by essentially the same procedure. However, PDEs involve several independent variables, so the calculations are typically lengthy. For the rest of this chapter, we merely outline the calculations, giving enough information to enable the reader to fill in the details.

In the previous example we solved the determining equations one at a time, using the terms multiplied by u_x^k before those multiplied by u_x^{k-1}. The information gained at each stage was then used to simplify the next equation. This is a very efficient technique that generalizes to higher-order PDEs (for which there may be many determining equations) as follows.

First write down the linearized symmetry condition, but do not expand each η^{J}. Instead, identify the terms in the linearized symmetry condition that are multiplied by the highest power of the highest-order derivative(s) of u. These terms yield some of the determining equations, which should now be solved. Then the results are used to simplify the remaining terms in the linearized symmetry condition. Now write down the terms that are multiplied by the highest remaining power of the highest remaining derivative(s), and solve the resulting determining equations. Iterate until the linearized symmetry condition has been completely satisfied.

This procedure generally works well, but sometimes the result is obtained more quickly by changing the order in which terms are used.

Example 8.3 The linearized symmetry condition for Burgers' equation,

$$u_t + uu_x = u_{xx}, \tag{8.52}$$

is

$$\eta^t + u\eta^x + u_x\eta = \eta^{xx} \qquad \text{when (8.52) holds.} \tag{8.53}$$

Once u_{xx} has been replaced by the left-hand side of (8.52), the highest-order derivative terms in (8.53) have a factor u_{xt}. We start by writing down those terms alone:

$$0 = -2\tau_x u_{xt} - 2\tau_u u_x u_{xt}.$$

This leads to

$$\tau_x = \tau_u = 0,$$

which removes many terms from the linearized symmetry condition; the remaining terms are

$$\begin{aligned}\eta_t - \xi_t u_x + (\eta_u - \tau_t)u_t - \xi_u u_x u_t + u\big(\eta_x + (\eta_u - \xi_x)u_x - \xi_u u_x^2\big) + u_x\eta \\ = \eta_{xx} + (2\eta_{xu} - \xi_{xx})u_x + (\eta_{uu} - 2\xi_{xu})u_x^2 - \xi_{uu}u_x^3 \\ + (\eta_u - 2\xi_x - 3\xi_u u_x)(u_t + uu_x).\end{aligned}$$

In particular, the terms multiplied by u_t are

$$(\eta_u - \tau_t)u_t - \xi_u u_x u_t = (\eta_u - 2\xi_x - 3\xi_u u_x)u_t.$$

This yields two determining equations:

$$\xi_u = 0, \qquad \xi_x = \tfrac{1}{2}\tau'(t).$$

Hence

$$\xi = \tfrac{1}{2}\tau'(t)x + \alpha(t),$$

for some function α. The remaining terms in the linearized symmetry condition determine α and τ up to five arbitrary constants. The Lie algebra of point symmetry generators is spanned by

$$X_1 = \partial_x, \qquad X_2 = \partial_t, \qquad X_3 = t\partial_x + \partial_u,$$

$$X_4 = x\partial_x + 2t\partial_t - u\partial_u, \qquad X_5 = xt\partial_x + t^2\partial_t + (x - ut)\partial_u.$$

N.B. In the above calculations, it was expedient to give terms multiplied by u_t precedence over those multiplied by powers of u_x. This is usual for *evolution equations*, which are PDEs of the form

$$u_t = F(x, t, u, u_x, u_{xx}, u_{xxx}, \ldots).$$

(F contains derivatives of u with respect to x only, not t.) For Burgers' equation, F has a term proportional to u_{xx}, so it is natural for u_t to take precedence over the u_x terms.

If a PDE is linear and homogeneous, it has an infinite-dimensional Lie algebra of point symmetry generators. The principle of linear superposition states that if $u(x, t)$ and $U(x, t)$ are solutions of the PDE, then so is

$$\hat{u} = u + \varepsilon U(x, t)$$

(for all ε). Therefore

$$X_U = U(x, t)\partial_u \tag{8.54}$$

is a symmetry generator, for any solution $U(x, t)$. The PDE has infinitely many linearly independent solutions, so the Lie algebra is infinite dimensional. Similarly, if u satisfies an inhomogeneous linear PDE and $U(x, t)$ is any solution of the related homogeneous PDE, then (8.54) is a symmetry generator. Suppose that a given nonlinear PDE has point symmetry generators that depend upon arbitrary solutions of some linear homogeneous equation. Then, by comparing the symmetry generators of the two equations, one may be able to linearize the original PDE. The aim is to construct a point transformation that maps the nonlinear PDE to the linear equation (or to a related inhomogeneous equation). The next example shows how this is done.

Example 8.4 The Lie algebra of point symmetry generators of the Thomas equation

$$u_{xt} = u_x u_t - 1 \tag{8.55}$$

is spanned by

$$\begin{gathered} X_1 = \partial_x, \qquad X_2 = \partial_t, \qquad X_3 = \partial_u, \qquad X_4 = x\partial_x - t\partial_t, \\ \{X_V = V(x, t)e^u \partial_u : V_{xt} = V\}. \end{gathered} \tag{8.56}$$

The fact that the Lie algebra of the Thomas equation depends upon solutions of

$$v_{xt} = v \tag{8.57}$$

suggests that we might be able to transform the Thomas equation into (8.57). The Lie algebra of point symmetry generators of (8.57) is spanned by

$$X_1 = \partial_x, \qquad X_2 = \partial_t, \qquad X_3 = v\partial_v, \qquad X_4 = x\partial_x - t\partial_t, \\ \{X_V = V(x,t)\partial_v : V_{xt} = V\}. \tag{8.58}$$

We compare the Lie algebras (8.56) and (8.58), with the aim of finding a point transformation that maps one to the other. Both Lie algebras have the same action on x and t, which suggests that these variables are unchanged by the point transformation. So we seek a change of variables $v = f(u)$ that maps (8.56) to (8.58). The usual change of variables formula for infinitesimal generators leads to

$$v = e^{-u}. \tag{8.59}$$

The reader can check that this transformation does indeed linearize the Thomas equation.

If the Lie algebra of point symmetry generators of a given PDE is finite dimensional, the PDE cannot be linearized by a point transformation. The point symmetry generators of the untransformed PDE are mapped to those of the transformed PDE (and vice versa) by the chain rule, so the dimension of the Lie algebra is unchanged by any point transformation. However, some PDEs with finite Lie algebras can be linearized by nonpoint transformations.

8.2 The Linearized Symmetry Condition for General PDEs

Everything in the previous section generalizes to PDEs with M dependent variables $u = (u^1, \dots, u^M)$ and N independent variables $x = (x^1, \dots, x^N)$. To keep the notation to a minimum, we shall use $u^{(k)}$ to denote the set of dependent variables and their partial derivatives of order k or less. By now the procedure for determining the linearized symmetry condition should be familiar; the only difference is that we need more indices! Therefore we describe only the main points, without going into detail.

Suppose that X is the infinitesimal generator of a one-parameter Lie group of point transformations, that is,

$$X = \xi^i(x,u)\partial_{x^i} + \eta_\alpha(x,u)\partial_{u^\alpha}. \tag{8.60}$$

The characteristic of the group is $\mathbf{Q} = (Q_1, \dots, Q_M)$, where

$$Q_\alpha = \eta_\alpha - \xi^i u^\alpha_{x^i}, \qquad \alpha = 1, \dots, M. \tag{8.61}$$

To prolong the linearized group action, we need the total derivatives,

$$D_{x^i} \equiv \partial_{x^i} + u^\alpha_{x^i}\partial_{u^\alpha} + \cdots.$$

The first prolongation of X is

$$X^{(1)} = X + \eta^l_\alpha\big(x, u^{(1)}\big)\partial_{u^\alpha_{x^l}},$$

where

$$\eta^l_\alpha\big(x, u^{(1)}\big) = D_{x^l} Q_\alpha + \xi^i D_{x^l} u^\alpha_{x^i}. \tag{8.62}$$

Similarly, the general formula for the prolonged generators is

$$X = \xi^i(x, u)\partial_{x^i} + \eta_\alpha(x, u)\partial_{u^\alpha} + \eta^{\mathrm{J}}_\alpha \partial_{u^\alpha_{\mathrm{J}}}, \tag{8.63}$$

where $u^\alpha_{\mathrm{J}} = D_{\mathrm{J}} u^\alpha$ and

$$\eta^{\mathrm{J}}_\alpha = D_{\mathrm{J}} Q_\alpha + \xi^i D_{\mathrm{J}} u^\alpha_{x^i}. \tag{8.64}$$

Here

$$D_{\mathrm{J}} = D^{j_1}_{x^1} D^{j_2}_{x^2} \ldots D^{j_M}_{x^M},$$

and the prolonged generator (8.63) is assumed to contain all terms that are needed to describe the linearized group action on a given PDE.

For simplicity, we shall consider only PDEs of the form

$$\Delta_\beta \equiv u_{\sigma_\beta} - \omega_\beta\big(x, u^{(n)}\big) = 0, \qquad \beta = 1, \ldots, M, \tag{8.65}$$

where each u_{σ_β} is a "highest derivative" of some u^α, in the sense that no other term in the system contains either u_{σ_β} or any of its derivatives. This enables us to replace each u_{σ_β} in the linearized symmetry condition by the corresponding ω_β. The resulting system can then be split into the determining equations by equating powers of the remaining derivatives of u.

N.B. We have restricted attention to systems of PDEs with the same number of equations as dependent variables. To avoid technical difficulties, we assume that the system is analytic and that there exists a unique solution to a Cauchy problem with arbitrary initial data in the region of interest. The beginner should not spend time investigating this technicality, as a firm grasp of basic symmetry methods is needed before its significance can be appreciated!

Example 8.5 Some PDEs have linearizing transformations that are not point transformations in the original variables but appear as point transformations when the PDE is written as a first-order system. We have shown that Burgers' equation,

$$u_t + uu_x = u_{xx}, \tag{8.66}$$

has a five-dimensional Lie algebra of point symmetry generators. Therefore it is not linearizable by a point transformation. Burgers' equation can be written as a conservation law, as follows:

$$u_t + \left(\tfrac{1}{2}u^2 - u_x\right)_x = 0.$$

Hence there exists a potential, v, such that

$$\begin{aligned} u_x &= v_t + \tfrac{1}{2}u^2, \\ v_x &= u. \end{aligned} \tag{8.67}$$

If we eliminate v from (8.67), we obtain Burgers' equation. Alternatively, we could eliminate u to obtain the "potential form" of Burgers' equation,

$$v_t + \tfrac{1}{2}v_x^2 = v_{xx}. \tag{8.68}$$

Let us seek the Lie point symmetries of the system (8.67). From (8.60), the Lie point symmetry generators are of the form

$$X = \xi(x, t, u, v)\partial_x + \tau(x, t, u, v)\partial_t + \eta(x, t, u, v)\partial_u + \chi(x, t, u, v)\partial_v.$$

(Here the variables and functions are assigned distinct letters to avoid a proliferation of indices.) The linearized symmetry condition is

$$\eta^x = \chi^t + u\eta, \qquad \chi^x = \eta, \qquad \text{when (8.67) holds.} \tag{8.69}$$

After a straightforward calculation (which is left as an exercise), we obtain an infinite-dimensional Lie algebra that is spanned by

$$\begin{gathered} X_1 = \partial_x, \qquad X_2 = \partial_t, \qquad X_3 = \partial_v, \qquad X_4 = x\partial_x + 2t\partial_t - u\partial_u, \\ X_5 = t\partial_x + \partial_u + x\partial_v, \qquad X_6 = xt\partial_x + t^2\partial_t + (x - tu)\partial_u + \left(\tfrac{1}{2}x^2 + t\right)\partial_v, \\ \left\{X_W = \left[W_x(x,t) + \tfrac{1}{2}W(x,t)u\right]e^{v/2}\partial_u + W(x,t)e^{v/2}\partial_v : W_t = W_{xx}\right\}. \end{gathered} \tag{8.70}$$

If we temporarily ignore all terms multiplied by ∂_v, we recover the five generators that were found in Example 8.3. We also obtain generators of the form

$$X_W = \left[W_x(x,t) + \tfrac{1}{2}W(x,t)u\right]e^{v/2}\partial_u, \tag{8.71}$$

which do not generate point symmetries of Burgers' equation. Point symmetry generators depend only on (x, t, u), whereas generators of the form (8.71) depend also on the potential, v. (They are called *potential symmetries* of Burgers' equation.)

Now we consider the restriction of the generators (8.70) to the variables (x, t, v), ignoring the ∂_u terms. The restricted generators are

$$\begin{gathered} X_1 = \partial_x, \qquad X_2 = \partial_t, \qquad X_3 = \partial_v, \qquad X_4 = x\partial_x + 2t\partial_t, \\ X_5 = t\partial_x + x\partial_v, \qquad X_6 = xt\partial_x + t^2\partial_t + \left(\tfrac{1}{2}x^2 + t\right)\partial_v, \\ \left\{X_W = W(x, t)e^{v/2}\partial_v : W_t = W_{xx}\right\}. \end{gathered} \tag{8.72}$$

These depend on x, t, and v, but not u; they are the point symmetry generators for the potential form of Burgers' equation. The generators X_W depend on arbitrary solutions of the heat equation

$$w_t = w_{xx},$$

and it is easily shown that (8.68) is mapped to the heat equation by writing

$$w = e^{-v/2}. \tag{8.73}$$

(The derivation of this result is left as an exercise.) Therefore

$$w_x = -\tfrac{1}{2}v_x e^{-v/2} = -\tfrac{1}{2}uw,$$

which leads to the well-known *Hopf–Cole transformation*:

$$u = -2\frac{w_x}{w}. \tag{8.74}$$

This is not a point transformation of Burgers' equation, but point symmetries of the system (8.67) have enabled us to derive it.

8.3 Finding Symmetries by Computer Algebra

The linearized symmetry condition gives us a fairly systematic approach to finding Lie point symmetries. However, as the number of variables or the order of the differential equation increases, the calculations rapidly become unmanageable. (The examples that we have studied so far are relatively simple, being low-order PDEs with few variables.) Over the past few years, symmetry-finding packages have been developed for many computer algebra systems. This section is a brief introduction to some packages that are currently available. It is not comprehensive but is intended to give the reader enough information to get started. Many packages have subroutines to enable the user to find exact solutions, calculate commutators, and so on. These useful tools vary greatly between packages, and we do not describe them here. Packages and detailed documentation can be downloaded from the websites listed at the end of the section.

To obtain the Lie point symmetries of a given PDE (8.65), a package must first calculate the determining equations by splitting the linearized symmetry condition. The user inputs the differential equation(s) and tells the program which terms u_{σ_β} are to be eliminated. Many packages require the PDE to be polynomial, at least in the derivatives of u, in order to be able to derive the determining equations by equating powers of the derivatives. PDEs that are rational polynomials, such as the nonlinear filtration equation

$$u_t = \frac{u_{xx}}{1 + u_x^2}, \tag{8.75}$$

can be input in polynomial form, for example,

$$\left(1 + u_x^2\right)u_t - u_{xx} = 0.$$

Usually the derivation of the determining equations is straightforward and is accomplished without intervention by the user. If the system of PDEs is very large, high order, or complicated, the package may run into memory limitations. One way around this problem is to start by finding the first few determining equations, solving them, and using the results to simplify the remaining calculations. This is the same procedure that we used earlier to find symmetries by hand.

The output from the above calculations should be a set of determining equations whose general solution yields all Lie point symmetry generators. For many PDEs, there are redundancies in the "raw" set of determining equations that one obtains by equating powers of derivatives. For example, suppose that the determining equations include

$$\xi_u = 0,$$

$$\xi_{uu} = 0.$$

The second of these is redundant, because it is a consequence of the first. Symmetry-finding packages normally produce a reduced list of determining equations, with the redundancies eliminated.

Having obtained some or all of the determining equations, the next step is to solve them. Many packages do this automatically, using basic integration techniques wherever appropriate. Some of these packages use the computer algebra system's general routines for solving differential equations; others have their own routines.

Warning: Some packages with automatic solvers occasionally fail to obtain the general solution of the determining equations. Therefore such packages may not find all of the symmetry generators. Surprisingly, it is possible to determine the size of the Lie algebra for a given PDE without first solving the determining equations. There are several programs that do this, enabling the user to check that a symmetry-finding package has obtained all symmetries.

Not all symmetry-finding packages incorporate differential equation solvers. Packages without solvers are designed so that the user is in control of the solution process. Such packages are usually very reliable and are capable of dealing with very large and complicated systems of PDEs. The determining equations are usually solved in stages, just as if the calculations were being done by hand. When a determining equation has been solved (or integrated to yield a lower-order equation), the result can be fed back into the remaining determining equations. The package may then be able to split some equations by equating powers of the dependent and independent variables. At each stage, the aim is to obtain whatever information can be deduced easily and to use this to simplify the unsolved equations.

The above strategy is based on the premise that at least one of the determining equations is easy to solve (or reduce to a lower-order PDE). What is to be done if the set of determining equations is highly coupled, so that the system seems to be intractable? Coupled systems of linear algebraic equations can be solved by Gaussian elimination, which reduces the problem to that of solving the system

$$A\mathbf{x} = \mathbf{b},$$

where the matrix A is in echelon (i.e., upper triangular) form. The reduced system consists of equations of increasing complexity (reading upwards). The simplest equation is easy to solve and yields one component of $\mathbf{x}$. This information enables us to solve for the next component, and so on. A similar strategy is used to solve coupled systems of polynomial nonlinear algebraic equations. The variables are ordered in an appropriate way, and Buchberger's algorithm is used to construct a reduced Gröbner basis. Roughly speaking, the Gröbner basis is a system of equations in "echelon" form, with the same set of solutions as the original system. The reduced system is generally much easier to solve than the original system.

The same approach has been adapted to linear and nonlinear systems of differential equations. Programs are available that use differentiation, multiplication, and addition to arrive at a "differential Gröbner basis." The outcome depends chiefly on the ordering of the variables. There are several standard ordering schemes that succeed in simplifying many systems. Even so, it is sometimes difficult to obtain a differential Gröbner basis (for reasons that are beyond the scope of this book).

The best way to learn more about symmetry-finding packages is to acquire one and carry out some experiments. Many different packages are available for the various computer algebra systems (CASs). Here is a selection of some widely used packages (one for each of three major CASs, and one that does not need a pre-installed CAS).

LIE and BIGLIE: Self-contained; for IBM-compatible PCs only.
Author: A. K. Head
http://archives.math.utk.edu/software/msdos/adv.diff.equations/lie51.html

LIE is written in muMATH, a DOS-based CAS that runs on IBM-compatible PCs. A limited version of muMATH is bundled with LIE, so the package is self-contained. Version 5.1 is available at the time of writing; unlike earlier versions, it seems to run safely in Windows 95 DOS.

LIE is able to determine symmetry generators for PDEs that are polynomial in the derivatives of u. Usually the package derives and solves the determining equations automatically, using a few basic integration techniques. The main constraint on LIE is that muMATH allows only 256 Kb of memory for the program and workspace. Nevertheless, LIE is able to deal with surprisingly complex systems, because it uses the available memory very efficiently. A related program, BIGLIE, is able to determine Lie point symmetries for PDEs that are too complex for LIE.

SPDE: Requires REDUCE.
Author: F. Schwarz
http://casun2.gmd.de/ Email: reduce-netlib@rand.org

The current version (1.0) of SPDE can determine symmetry generators for PDEs that are algebraic in their arguments. The package is able to obtain the size of the symmetry group directly from the determining equations. For this to be accomplished, a differential Gröbner basis is constructed. SPDE integrates the reduced determining eqations automatically; it is guaranteed to find all generators with algebraic coefficients for nonlinear PDEs of order 2 or more.

SPDE can be used interactively at the above website; the full package is available by email.

SYMMGRP.MAX: Requires MACSYMA.
Authors: B. Champagne, P. Winternitz, and W. Hereman
ftp://mines.edu/pub/papers/math_cs_dept/software/symmetry/

SYMMGRP.MAX is a flexible, well-tested package that is designed to be used interactively. It can cope with very large systems, calculating a few determining equations at a time. The user has to solve one or more of these, the results of which simplify the next few determining equations, and so on. SYMMGRP.MAX is most effective when it is used in conjunction with a package that constructs differential Gröbner bases, such as DIFFGROB2 (see below).

DIFFGROB2: Requires Maple.
Author: E. L. Mansfield
ftp://ftp.ukc.ac.uk/pub/maths/liz/

DIFFGROB2 is an effective, powerful package for simplifying systems of linear or nonlinear PDEs. It can construct differential Gröbner bases for general linear systems and some nonlinear systems. The choice of term ordering is crucial, so the package may be used interactively to enable the user to try out various orderings. DIFFGROB2 is a very useful adjunct to symmetry-finding packages that do not include Gröbner basis algorithms (e.g., SYMMGRP.MAX). It is virtually a necessity if one seeks "nonclassical" symmetries, whose determining equations are nonlinear (see §9.3).

Further Reading

Many important nonlinear PDEs admit linearizing transformations, which can often be found by symmetry methods. This chapter has introduced such transformations and has shown how they may be constructed. Nevertheless, our treatment is neither rigorous nor exhaustive. To find out more, the reader should consult Chapter 6 of Bluman and Kumei (1989). These authors also present a thorough discussion of potential symmetries, with many examples and applications.

The Cauchy–Kovalevskaya theorem gives conditions for the existence of solutions to the Cauchy problem for an analytic system of PDEs. Olver (1993) explains why the existence and uniqueness of solutions ensures that the linearized symmetry condition yields the most general (connected) symmetry group.

Readers intending to use SPDE or DIFFGROB2 should first become familiar with Gröbner bases for polynomial nonlinear systems. The text by Cox, Little, and O'Shea (1992) contains a very readable account of Buchberger's algorithm and is intended for newcomers to commutative algebra. The simplest introduction to DIFFGROB2 is by Mansfield and Clarkson (1997), who describe various strategies for simplifying the determining equations.

Readers who wish to know more about symmetry software should consult the paper by Hereman (1996). This clear, careful review includes details of many packages other than those listed in §8.3.

Exercises

8.1 Show that the hodograph transformation $(\hat{x}, \hat{t}, \hat{u}) = (u, t, x)$ is a symmetry of the nonlinear filtration equation (8.75).

8.2 Determine the Lie point symmetries of

$$u_t = u_x^3.$$

8.3 Derive the Lie point symmetry generators for the heat equation,

$$u_t = u_{xx}.$$

8.4 Show that the Lie algebra of point symmetry generators of (8.68) is spanned by (8.72).

8.5 Compare the symmetry generators of the heat equation with (8.72), and hence derive the transformation (8.73).

8.6 In shallow water, one-dimensional motion of long waves is governed by the system

$$u_t + uu_x + v_x = 0,$$
$$v_t + uv_x + vu_x = 0.$$

Find the Lie point symmetry generators for this system, and hence obtain a linearizing transformation.

8.7 Obtain a symmetry-finding computer algebra package and use it to rederive the symmetries that have been obtained in this chapter.

9

Methods for Obtaining Exact Solutions of PDEs

If he's a change, give me a constancy.

(Charles Dickens: Dombey and Son)

9.1 Group-Invariant Solutions

Armed with the methods from the previous chapter, we are able to find the Lie point symmetries of a given PDE systematically. This chapter describes how to use symmetries to construct exact solutions. The methods used are a generalization of §4.3, which the reader may wish to revisit before continuing.

Nearly all exact methods relate a given PDE to one or more ODEs. For example, the general solution of a first-order quasilinear PDE is constructed by integrating the characteristic equations. For most PDEs, we cannot write down the "general solution" but have to rely on various ansätze. We may seek similarity solutions, travelling waves, separable solutions, and so on. Many of these methods involve nothing more than looking for solutions that are invariant under a particular group of symmetries. For example, PDEs for $u(x,t)$ whose symmetry generators include

$$X_1 = \partial_x, \qquad X_2 = \partial_t$$

generally have travelling wave solutions of the form

$$u = F(x - ct). \tag{9.1}$$

These solutions are invariant under the group generated by

$$X = cX_1 + X_2 = c\partial_x + \partial_t, \tag{9.2}$$

because both u and $x - ct$ are invariants. In the same way, PDEs with scaling symmetries admit similarity solutions, which are constructed from the invariants of the group.

This idea is easily generalized to any Lie group of symmetries of a given PDE

$$\Delta = 0. \tag{9.3}$$

For now, let us restrict attention to scalar PDEs with two independent variables. Recall that a solution $u = u(x, t)$ is invariant under the group generated by

$$X = \xi \partial_x + \tau \partial_t + \eta \partial_u$$

if and only if the characteristic vanishes on the solution. In other words, every invariant solution satisfies the invariant surface condition

$$Q \equiv \eta - \xi u_x - \tau u_t = 0. \tag{9.4}$$

Usually (9.4) is much easier to solve than the original PDE! Having solved (9.4), we can find out which solutions also satisfy (9.3). For example, the group generated by (9.2) has the characteristic

$$Q = -cu_x - u_t. \tag{9.5}$$

The travelling wave ansatz (9.1) is the general solution of the invariant surface condition $Q = 0$.

For now, suppose that ξ and τ are not both zero. Then the invariant surface condition is a first-order quasilinear PDE that can be solved by the method of characteristics. The characteristic equations are

$$\frac{dx}{\xi} = \frac{dt}{\tau} = \frac{du}{\eta}. \tag{9.6}$$

If $r(x, t, u)$ and $v(x, t, u)$ are two functionally independent first integrals of (9.6), every invariant of the group is a function of r and v. Usually, it is convenient to let one invariant play the role of a dependent variable. Suppose (without loss of generality) that $v_u \neq 0$; then the general solution of the invariant surface condition is

$$v = F(r). \tag{9.7}$$

This solution is now substituted into the PDE (9.3) to determine the function F.

If r and v both depend on u, it is necessary to find out whether the PDE has any solutions of the form

$$r = c. \tag{9.8}$$

These are the only solutions of the invariant surface condition that are not (locally) of the form (9.7). If r is a function of the independent variables x and t only, then (9.8) cannot yield a solution $u = u(x, t)$.

Example 9.1 Among the many symmetries of the heat equation,

$$u_t = u_{xx}, \tag{9.9}$$

there is a two-parameter Lie group of scalings, which is generated by

$$X_1 = x\partial_x + 2t\partial_t, \qquad X_2 = u\partial_u.$$

Every generator of a one-parameter Lie group of scalings is of the form

$$X = hX_1 + kX_2,$$

for some constants h, k. Remember that if λ is any nonzero constant, X and λX generate the same one-parameter group. (The group parameter ε is changed, but this does not affect the group.) Therefore if $h \neq 0$ we may assume that $h = 1$ without loss of generality; if $h = 0$, we set $k = 1$.

Suppose that $h = 1$, so that

$$X = x\partial_x + 2t\partial_t + ku\partial_u. \tag{9.10}$$

The invariant surface condition is

$$Q \equiv ku - xu_x - 2tu_t = 0,$$

which is solved by integrating the characteristic equations

$$\frac{dx}{x} = \frac{dt}{2t} = \frac{du}{ku}.$$

Simple quadrature yields the invariants

$$r = xt^{-1/2}, \qquad v = ut^{-k/2}.$$

Because r is independent of u, every invariant solution is of the form

$$v = F(r),$$

which is equivalent to

$$u = t^{k/2}F(xt^{-1/2}). \tag{9.11}$$

Differentiating (9.11), we obtain

$$u_t = t^{(k-2)/2}\left(-\tfrac{1}{2}rF'(r) + \tfrac{1}{2}kF(r)\right),$$
$$u_{xx} = t^{(k-2)/2}F''(r).$$

Therefore (9.11) is a solution of the heat equation if

$$F'' + \tfrac{1}{2}rF' - \tfrac{1}{2}kF = 0. \tag{9.12}$$

The general solution of (9.12) is

$$F(r) = c_1U\left(k + \tfrac{1}{2},\, 2^{-1/2}r\right) + c_2V\left(k + \tfrac{1}{2},\, 2^{-1/2}r\right),$$

where $U(p, z)$ and $V(p, z)$ are parabolic cylinder functions. If k is an integer, these functions can be expressed in terms of elementary functions and their integrals. For example, if $k = 0$ then

$$F(r) = c_1 \operatorname{erf}\left(\frac{r}{2}\right) + c_2,$$

where

$$\operatorname{erf}(z) = \frac{2}{\sqrt{\pi}}\int_0^z e^{-\zeta^2}\,d\zeta$$

is the error function. If $k = -1$,

$$F = c_1e^{-r^2/4} + c_2e^{-r^2/4}\int_0^r e^{\zeta^2/4}\,d\zeta.$$

Substituting these results into (9.11), we obtain a large family of solutions which includes the fundamental solution

$$u = t^{-1/2}e^{-x^2/4t}, \qquad (k = -1),$$

the error function solution

$$u = \operatorname{erf}\left(\frac{x}{2\sqrt{t}}\right), \qquad (k = 0),$$

and many other well-known solutions.

So far, we have assumed that ξ and τ are not both zero, so that at least one of the invariants r, v depends on u. If ξ and τ are zero at some point on an invariant surface, the invariant surface condition (9.4) implies that η is also zero at that point. Hence the generator X vanishes there, and the point is invariant. If ξ and τ are identically zero, the only possible invariant solutions are those for which

$$\eta(x, t, u) \equiv 0. \tag{9.13}$$

Such solutions consist entirely of points that are invariant under the group generated by X. Of course, (9.13) need not have any solutions of the form $u = u(x, t)$. For instance, if

$$X = \partial_u$$

then

$$(\xi, \tau, \eta) = (0, 0, 1),$$

so (9.13) cannot be satisfied. There are other generators $X = \eta\partial_u$ for which (9.13) does have solutions $u = u(x, t)$. It is usually easy to check whether any of these solutions also solve the PDE (9.3).

Example 9.2 In the previous example, we did not look for scaling-invariant solutions generated by

$$X = u\partial_u.$$

Here ξ and τ are both zero, but the invariant surface condition does have a solution, namely

$$u = 0.$$

This is a solution of the heat equation, even though it is not a particularly interesting one!

Although travelling waves and scale-invariant (similarity) solutions will be familiar to most readers, there is nothing that sets these transformations apart from other symmetries. The procedure for finding group-invariant solutions is the same, whatever the group is.

Example 9.3 The nonlinear filtration equation

$$u_t = \frac{u_{xx}}{1 + u_x^2} \tag{9.14}$$

has the five-parameter Lie group of point symmetries generated by

$$\begin{aligned} X_1 &= \partial_x, \qquad X_2 = \partial_t, \qquad X_3 = \partial_u, \\ X_4 &= x\partial_x + 2t\partial_t + u\partial_u, \qquad X_5 = u\partial_x - x\partial_u. \end{aligned} \tag{9.15}$$

The one-parameter group of rotations generated by X_5 has invariants

$$r = t, \qquad v = \sqrt{x^2 + u^2}.$$

Any invariant solution $v = F(r)$ is equivalent to

$$u = \pm\sqrt{F(t)^2 - x^2}. \tag{9.16}$$

Substituting (9.16) into the PDE (9.14) leads to the ODE

$$F'(r) = -\frac{1}{F(r)}. \tag{9.17}$$

[Readers may be surprised that the second-order PDE has been reduced to a *first-order* ODE. This occurs because (9.14) has no second-order derivatives with respect to the invariant coordinate t.] The general solution of (9.17) is

$$F(r) = \pm\sqrt{c_1 - 2r},$$

so the solutions of (9.14) that are invariant under the rotations generated by X_5 are

$$u = \pm\sqrt{c_1 - 2t - x^2}. \tag{9.18}$$

The idea of looking for group-invariant solutions generalizes quite naturally to PDEs with any number of independent and dependent variables. In general, a one-parameter group that acts nontrivially on one or more independent variables can be used to reduce the number of independent variables by one. For example, suppose that a scalar PDE with three independent variables is written in terms of invariants of a one-parameter group. There are three functionally independent invariants, at least one of which depends on u, so the original PDE reduces to a scalar PDE with two independent variables. A further one-parameter symmetry group is needed to reduce this PDE to an ODE.

For systems of PDEs with M dependent variables, $u = (u_1, \ldots, u_M)$, the characteristic $\mathbf{Q}$ has M components, each of which is zero on any invariant solution. For example, if there are two independent variables x and t, the solution $u = u(x, t)$ is invariant if and only if

$$Q_\alpha\big|_{u=u(x,t)} \equiv X\left[u^\alpha - u^\alpha(x, t)\right]\big|_{u=u(x,t)} = 0, \qquad \alpha = 1, \ldots, M.$$

Just as for scalar PDEs, the best strategy is to start by solving the invariant surface condition

$$Q_\alpha = 0, \qquad \alpha = 1, \ldots, M; \tag{9.19}$$

then substitute the solution into the given system of PDEs to obtain a reduced problem.

Example 9.4 The following system models two-dimensional free convection of an incompressible viscous fluid between heated horizontal boundaries:

$$\begin{aligned} uu_x + vu_y &= u_{xx} + u_{yy} - p_x, \\ uv_x + vv_y &= v_{xx} + v_{yy} - p_y + \lambda\theta, \\ u\theta_x + v\theta_y &= \sigma(\theta_{xx} + \theta_{yy}), \\ u_x + v_y &= 0. \end{aligned} \tag{9.20}$$

(This model uses the Boussinesq approximation. The fluid velocity has components u, v in the horizontal and vertical directions, respectively; p is the pressure; θ is the temperature perturbation. All variables are suitably nondimensionalized; the dimensionless parameters are the Grashof number, λ, and the Prandtl number, σ.)

The system (9.20) has the five-parameter Lie group of point symmetries generated by

$$\begin{gathered} X_1 = \partial_x, \qquad X_2 = \partial_y, \qquad X_3 = \partial_p, \qquad X_4 = \lambda y \partial_p + \partial_\theta, \\ X_5 = x\partial_x + y\partial_y - u\partial_u - v\partial_v - 2p\partial_p - 3\theta\partial_\theta. \end{gathered} \tag{9.21}$$

The one-parameter groups generated by X_1, X_2, and X_3 consist of translations, whereas X_5 generates scalings (which lead to similarity solutions). For X_4, the characteristic

$$\mathbf{Q} = (Q_u, Q_v, Q_p, Q_\theta)$$

has the components

$$Q_u = 0, \qquad Q_v = 0, \qquad Q_p = \lambda y, \qquad Q_\theta = 1.$$

Therefore no solutions are invariant under the group generated by X_4. However, we can use X_4 to obtain invariant solutions by combining it with X_1. The characteristic of the group generated by

$$X = X_1 + kX_4 = \partial_x + k\lambda y \partial_p + k\partial_\theta, \qquad k \neq 0 \tag{9.22}$$

has the components

$$Q_u = -u_x, \qquad Q_v = -v_x, \qquad Q_p = k\lambda y - p_x, \qquad Q_\theta = k - \theta_x. \tag{9.23}$$

Therefore the general solution of the invariant surface condition $\mathbf{Q} = 0$ is

$$\begin{aligned} u &= F(y), \\ v &= G(y), \\ p - k\lambda xy &= H(y), \\ \theta - kx &= K(y). \end{aligned} \tag{9.24}$$

Substituting (9.24) into the original system of PDEs (9.20) leads to the following system of ODEs:

$$\begin{aligned} GF' &= F'' - k\lambda y, \\ GG' &= G'' - H' + \lambda K, \\ kF + GK' &= \sigma K'', \\ G' &= 0. \end{aligned}$$

The last of these is easily solved:

$$G = c_1.$$

The remaining ODEs reduce to the linear system

$$\begin{aligned} F'' - c_1 F' &= k\lambda y, \\ K'' - \frac{c_1}{\sigma} K' &= \frac{k}{\sigma} F, \\ H' &= \lambda K. \end{aligned} \tag{9.25}$$

This system has elementary solutions containing six arbitrary constants. We have not considered boundary conditions, which constrain the solution set. For example, the above solution has $v = G = c_1$. If the boundaries are impermeable then v is zero there; our invariant solutions satisfy this condition if and only if $c_1 = 0$.

9.2 New Solutions from Known Ones

Suppose that we have found a solution that is invariant under a group of Lie point symmetries. If there are other symmetries under which this solution is not invariant, we can use them to map it to a family of new solutions. The procedure is exactly the same as for ODEs (see §4.3), as the next example shows.

Example 9.5 Earlier, we found that the nonlinear filtration equation

$$u_t = \frac{u_{xx}}{1 + u_x^2} \tag{9.26}$$

has a one-parameter family of solutions,

$$u = \sqrt{c_1 - 2t - x^2}, \tag{9.27}$$

each of which is invariant under $X_5 = u\partial_x - x\partial_u$. Let us try to construct new solutions from (9.27), using all remaining symmetry generators (9.15). The symmetries generated by $X_1 = \partial_x$ are

$$(\hat{x}, \hat{t}, \hat{u}) = (x + \varepsilon, t, u).$$

Therefore (9.27) is equivalent to

$$\hat{u} = \sqrt{c_1 - 2\hat{t} - (\hat{x} - \varepsilon)^2}.$$

Removing carets, we conclude that the group generated by X_1 maps (9.27) to a two-parameter family of solutions:

$$u = \sqrt{c_1 - 2t + (x - \varepsilon)^2}. \tag{9.28}$$

The one-parameter group generated by $X_3 = \partial_u$ is

$$(\hat{x}, \hat{t}, \hat{u}) = (x, t, u + \delta),$$

which maps (9.28) to the three-parameter family

$$u = \sqrt{c_1 - 2t + (x - \varepsilon)^2} + \delta. \tag{9.29}$$

Lie point symmetries generated by X_2 and X_4 produce no further solutions from (9.29); their action merely changes the values of the arbitrary constants c_1, δ, and ε. Therefore we can obtain no further solutions from (9.27) by the above method.

N.B. Each solution (9.29) is invariant under the one-parameter group generated by

$$\begin{aligned}\tilde{X}_5 &= (u - \delta)\partial_x - (x - \varepsilon)\partial_u \\ &= X_5 - \delta X_1 + \varepsilon X_3.\end{aligned} \tag{9.30}$$

Clearly, $\tilde{X}_5$ and X_5 are related. Both generate rotations in the (x, u) plane, but $\tilde{X}$ generates rotations about (ε, δ) rather than $(0, 0)$.

Given any solution that is invariant under a one-parameter symmetry group, we can derive a family of solutions by applying the remaining symmetries. (Of course, if the solution is invariant under all symmetries, it is the only member of the family.) In Chapter 10 we shall show that each of these solutions is invariant under (at least) a one-parameter symmetry group. Hence the set of all invariant solutions splits into equivalence classes; solutions are in the same class if they can be mapped to one another by a point symmetry. Typically there are only a few classes, which makes it fairly easy to classify all invariant solutions.

For homogeneous linear PDEs, there is another way to generate new solutions. For simplicity, we consider only scalar PDEs for $u(x,t)$. It is possible to show (from the linearized symmetry condition) that every generator of Lie point symmetries of a homogeneous linear PDE, $\Delta = 0$, is of the form

$$X = \xi(x,t)\partial_x + \tau(x,t)\partial_t + \big(f(x,t)u + U(x,t)\big)\partial_u, \tag{9.31}$$

where $u = U(x,t)$ is an arbitrary solution of $\Delta = 0$. The Lie algebra $\mathcal{L}$ of point symmetry generators splits into a finite-dimensional subalgebra $\mathcal{L}_0$, consisting of generators (9.31) with $U = 0$, and the infinite-dimensional abelian subalgebra $\mathcal{L}_\infty$ spanned by the generators

$$X_U = U(x,t)\partial_u, \qquad \text{where} \quad \Delta|_{u=U(x,t)} = 0. \tag{9.32}$$

Given any solution $u = U(x,t)$ of the PDE $\Delta = 0$, construct the generator $X_U \in \mathcal{L}_\infty$. Now calculate the commutator of X_U with each $X_i \in \mathcal{L}_0$ in turn. Each X_i is of the form

$$X_i = \xi_i(x,t)\partial_x + \tau_i(x,t)\partial_t + f_i(x,t)u\partial_u, \tag{9.33}$$

and therefore

$$[X_U, X_i] = \tilde{U}_i(x,t)\partial_u, \tag{9.34}$$

where

$$\tilde{U}_i(x,t) = f_i(x,t)U(x,t) - \xi_i(x,t)U_x(x,t) - \tau_i(x,t)U_t(x,t). \tag{9.35}$$

The commutator of any two generators is itself a symmetry generator, that is,

$$\tilde{U}_i(x,t)\partial_u \in \mathcal{L}.$$

Moreover, this generator is of the form (9.32), so

$$\tilde{U}_i(x,t)\partial_u \in \mathcal{L}_\infty.$$

Hence $u = \tilde{U}_i(x,t)$ is a solution of the PDE $\Delta = 0$. If new solutions are obtained, each one can be used in place of U in an attempt to create further new solutions by the above method.

Example 9.6 Recall from Example 8.4 that the Lie point symmetries of

$$u_{xt} = u \tag{9.36}$$

are generated by

$$X_1 = \partial_x, \qquad X_2 = \partial_t, \qquad X_3 = u\partial_u, \qquad X_4 = x\partial_x - t\partial_t, \tag{9.37}$$

and

$$\{X_U = U(x,t)\partial_u : U_{xt} = U\}. \tag{9.38}$$

Clearly the generators (9.37) form a basis for the finite-dimensional subalgebra $\mathcal{L}_0$, whereas (9.38) spans $\mathcal{L}_\infty$. Given any solution $u = U(x,t)$, the generators (9.37) produce

$$\begin{aligned} \tilde{U}_1 &= -U_x, \\ \tilde{U}_2 &= -U_t, \\ \tilde{U}_3 &= U, \\ \tilde{U}_4 &= tU_t - xU_x. \end{aligned} \tag{9.39}$$

Any of these solutions except $\tilde{U}_3$ may be (but need not be) different from U. For example, let us start with the travelling wave solution

$$U = e^{x+t} \tag{9.40}$$

(which is invariant under the group generated by $X_2 - X_1$). Neither $\tilde{U}_1$ nor $\tilde{U}_2$ is independent of U, but

$$\tilde{U}_4 = (t-x)e^{x+t} \tag{9.41}$$

is new. Repeating the process, now with $U = (t-x)e^{x+t}$, gives

$$\tilde{U}_4 = \left[(t-x)^2 + t + x\right]e^{x+t}. \tag{9.42}$$

It appears that X_1 produces a new solution, namely

$$\tilde{U}_1 = (t-x-1)e^{x+t},$$

but this is merely a linear superposition of (9.40) and (9.41), so we gain nothing new. However X_4 repeatedly produces new solutions; the next in the series is

$$\tilde{U}_4 = \left((t-x)^3 + 3(t^2 - x^2) + t - x\right)e^{x+t}. \tag{9.43}$$

If we had chosen a different starting solution, X_1 or X_2 might have been able to generate new solutions.

The above idea also works for nonlinear PDEs that are linearizable by an invertible transformation. Each solution of the nonlinear PDE corresponds to a solution of the linear equation. Therefore, one can use the above method to generate new solutions of the linear equation and then transform these back into solutions of the original nonlinear PDE. For instance, the Thomas equation

$$u_{xt} = u_x u_t - 1$$

can be linearized to (9.36), as shown in Example 8.4. The reader may verify that each of the new solutions (9.41)–(9.43) corresponds to a solution of the Thomas equation.

9.3 Nonclassical Symmetries

This section introduces a class of point transformation groups that are not symmetries but that can lead to exact solutions of a given PDE. To avoid undue complexity, we restrict attention to scalar PDEs for $u(x, t)$. The method generalizes to arbitrary PDEs, but the complexity of the calculations increases greatly with the number of variables.

Lie point symmetries of a given PDE map the set of all solutions to itself. This is useful as soon as we know some solutions, but it does not enable us to construct new solutions from nothing. To do that, we seek solutions that are invariant under a one-parameter Lie group of point transformations. All solutions that are invariant under the group with characteristic Q satisfy the PDE *and* the invariant surface condition; they are the solutions of the system

$$\begin{aligned} \Delta &= 0, \\ Q &= 0. \end{aligned} \tag{9.44}$$

If we seek only invariant solutions, it is worthwhile trying to determine all X that generate point symmetries of the system (9.44). The difficulty is that the second equation ($Q = 0$) depends on X! Nevertheless it is often possible to find all such generators (systematically). For some PDEs these lead to new invariant solutions that cannot be found from the "classical" point symmetries.

The linearized symmetry condition for systems states that the generator X corresponding to Q generates symmetries of the nth order PDE (9.44) if

$$X^{(n)}\Delta = 0, \qquad X^{(1)}Q = 0, \qquad \text{when (9.44) holds.} \tag{9.45}$$

(For now, we show the order of prolongation explicitly.) This condition is simplified by the identity

$$X^{(1)}Q = QQ_u \tag{9.46}$$

(which the reader is asked to derive as an exercise). From (9.46), it follows that

$$X^{(1)}Q = 0 \quad \text{when} \quad Q = 0.$$

Therefore X is a symmetry of (9.44) if

$$X^{(n)}\Delta = 0 \quad \text{when} \quad \Delta = 0 \quad \text{and} \quad Q = 0. \tag{9.47}$$

In other words, the group generated by X needs only to map the invariant solutions to themselves. The remaining solutions of $\Delta = 0$ need not be mapped to solutions.

Clearly, if X generates symmetries of the PDE, it satisfies the linearized symmetry condition

$$X^{(n)}\Delta = 0 \quad \text{when} \quad \Delta = 0, \tag{9.48}$$

and hence it satisfies the weaker condition (9.47). However, there may be generators X that satisfy (9.47) but not the linearized symmetry condition (9.48). These are generators of *nonclassical symmetries*.

The solutions of (9.47) are obtained by calculating $X^{(n)}\Delta$, eliminating some derivatives of u by using $\Delta = 0$ and $Q = 0$, and finally equating powers of the remaining derivatives of u to obtain the determining equations. This sounds like the same procedure that is used to find the classical symmetries, but there is an important difference. The determining equations are nonlinear, because $Q = 0$ involves the unknown functions (ξ, τ, η). Consequently, the set of nonclassical symmetry generators is not a vector space, let alone a Lie algebra. Furthermore, the determining equations are usually too hard to solve without computer algebra.

Some simplification is achieved by the observation that if X generates a nonclassical symmetry then so does λX, for any nonzero function λ. To derive this result, we use the identity $(\lambda X)^{(k)} = \lambda X^{(k)}$ when $Q = 0$, which is a consequence of the prolongation formula (8.25). If X generates nonclassical symmetries then

$$(\lambda X)^{(n)}\Delta\Big|_{\substack{\lambda Q=0\\ \Delta=0}} = \lambda X^{(n)}\Delta\Big|_{\substack{Q=0\\ \Delta=0}} = 0,$$

so λX also generates nonclassical symmetries. Therefore, without loss of generality, we set $\tau = 1$ if τ is nonzero and $\xi = 1$ if $\tau = 0$. (If both ξ and τ are zero, the invariant surface condition requires that η is also zero. Then there is no way of systematically finding the invariant surfaces unless a classical symmetry generator is known.)

Example 9.7 The Huxley equation,

$$u_t = u_{xx} + 2u^2(1-u), \tag{9.49}$$

has a two-parameter Lie group of (classical) point symmetries, generated by

$$X_1 = \partial_x, \qquad X_2 = \partial_t.$$

All solutions that are invariant under the group generated by X_1 are spatially uniform, so $u = F(t)$ where

$$F' = 2F^2(1-F). \tag{9.50}$$

The only other solutions that are invariant under classical symmetries are travelling (or stationary) waves, $u = F(x - ct)$, where

$$F'' + cF' + 2F^2(1-F) = 0. \tag{9.51}$$

Let us look for nonclassical symmetries with $\tau = 1$, so that the invariant surface condition amounts to the constraint

$$u_t = \eta - \xi u_x. \tag{9.52}$$

The linearized symmetry condition for these nonclassical symmetries is

$$\eta^{xx} - \eta^t + (4u - 6u^2)\eta = 0, \qquad \text{when (9.49) and (9.52) hold.} \tag{9.53}$$

The determining equations are obtained by writing (9.53) out in full, using (9.49) and (9.52) to eliminate u_{xx} and u_t, and then splitting the resulting equation by equating powers of u_x. This produces the system

$$\begin{aligned} \xi_{uu} &= 0, \\ \eta_{uu} - 2\xi_{xu} + 2\xi\xi_u &= 0, \\ 2\eta_{xu} - \xi_{xx} - (2\eta + 6u^3 - 6u^2)\xi_u + 2\xi\xi_x + \xi_t &= 0, \\ \eta_{xx} - 2\eta\xi_x + 2u^2(u-1)(\eta_u - 2\xi_x) - \eta_t + (4u - 6u^2)\eta &= 0. \end{aligned} \tag{9.54}$$

Some of these determining equations are nonlinear, but (9.54) is quite easily solved (because it happens to be in a "triangular" form). The general solution of the first equation is

$$\xi = A(x,t)u + B(x,t),$$

which leads to the solution of the second equation:

$$\eta = -\tfrac{1}{3}A^2u^3 + (A_x - AB)u^2 + C(x,t)u + D(x,t).$$

These results enable us to split the remaining equations by equating powers of u, and so on. As this is now a familiar process, we skip the details and simply present the result. The solutions of (9.54) are

$$\xi = 1, \qquad \eta = 0, \tag{9.55}$$

and

$$\xi = \pm(3u-1), \qquad \eta = 3u^2(1-u). \tag{9.56}$$

We already know about (9.55), which corresponds to the classical symmetry generator $X_1 + X_2$ (remember, $\tau = 1$). However (9.56) is new; it gives the nonclassical symmetry generators

$$X = \pm(3u-1)\partial_x + \partial_t + 3u^2(1-u)\partial_u. \tag{9.57}$$

The invariant surface condition for the nonclassical symmetries is

$$u_t \pm (3u-1)u_x = 3u^2(1-u), \tag{9.58}$$

which is easily solved by the method of characteristics. Two functionally independent invariants are

$$r = \left(\frac{1}{u} - 1\right)e^{t\pm x}, \qquad v = \frac{1}{u} + 2t \mp x.$$

Now we substitute $v = F(r)$ into the Huxley equation, which reduces to

$$F'' = 0.$$

Therefore $F(r) = c_1 r + c_2$; writing this in terms of the original variables, we obtain

$$u = \frac{1 - c_1 e^{t\pm x}}{2t \pm x - c_1 e^{t\pm x} + c_2}. \tag{9.59}$$

The solutions with $c_1 \neq 0$ are not obtainable by any classical reduction. If $c_1 = 0$, the solution $v = c_2$ is a travelling wave. There is also a travelling wave solution $r = c_3$.

We have not yet looked for nonclassical symmetries with $\tau = 0$, for which

$$X = \partial_x + \eta(x, t, u)\partial_u.$$

The invariant surface condition is

$$u_x = \eta, \tag{9.60}$$

so any invariant solution of the Huxley equation satisfies

$$u_t = \eta_x + \eta\eta_u + 2u^2(1 - u). \tag{9.61}$$

Without going into details, we find that the nonclassical linearized symmetry condition is

$$\eta_{xx} + 2\eta\eta_{xu} + \eta^2\eta_{uu} - 2u^2(1 - u)\eta_u - \eta_t + (4u - 6u^2)\eta = 0,$$

which is simply the integrability condition for the system (9.60), (9.61). There is only one equation, because all derivatives of u are eliminated when (9.60) and (9.61) are taken into account. Consequently, we are unable to proceed further, except by trying various ansätze.

The above example is atypical in that the determining equations (with $\tau = 1$) are easy to solve! For most PDEs with nonclassical symmetries, the determining equations have to be simplified with the aid of computer algebra. The package DIFFGROB2 (discussed in Chapter 8) is able to deal with many nonlinear systems of overdetermined PDEs; nevertheless, it does not always succeed.

For many PDEs, every nonclassical reduction is also obtainable with classical symmetries. However, a few PDEs (such as the Boussinesq equation) have large families of solutions that cannot be found with classical symmetries. The reason for this is not yet understood.

Further Reading

The symmetries, linearizing transformations, and invariant solutions of many well-known PDEs have been classified during the past three decades. Most of these results are included in the useful handbook by Ibragimov (1994, 1995).

We have not taken initial or boundary conditions into account. Roughly speaking, an invariant solution to a boundary value problem exists if the PDE, the domain, and the boundary conditions are all invariant under the symmetry group. For example, Poiseuille flow in a cylindrical pipe is possible because the equations of motion, the domain (a cylinder), and the boundary conditions (zero flow at the pipe wall) are invariant under rotations about the pipe's axis. (N.B. The existence of an invariant solution is no guarantee of its stability. For example, turbulent pipe flow is not axisymmetric.) Under some circumstances, it is possible to use invariant solutions of the PDE to build up a composite solution that satisfies given boundary conditions. Bluman and Kumei (1989) describe how to do this for linear PDEs.

In many applications, invariant solutions describe the limiting behaviour of a system, "far away" from initial or boundary conditions. For instance, the limiting behaviour of a parabolic PDE on an unbounded domain is often described by a similarity solution. Barenblatt (1996) describes many physical problems for which this occurs and develops techniques for the analysis of scale-invariant limiting behaviour.

Nonclassical symmetries have attracted much research effort recently, as reliable computer algebra packages have become available. The analysis of the Boussinesq equation by Clarkson (1996) is a good starting point for readers who wish to find out more about nonclassical symmetries.

Exercises

9.1 In Chapter 8, we showed that the Thomas equation (8.55) may be linearized to

$$u_{xt} = u,$$

whose symmetries include the scalings generated by

$$X = x\partial_x - t\partial_t + ku\partial_u.$$

Find all solutions of the above PDE that are invariant under the group generated by X (for arbitrary k). What are the solutions if $k = \frac{1}{2}$? Use the methods of §9.2 to construct a large family of invariant solutions from the ones that you have already found. Now use your results to construct a family of solutions of the Thomas equation. [Hint: Use $r = \sqrt{xt}$ as one invariant.]

9.2 Show that the characteristic of a one-parameter Lie group of point symmetries of a scalar PDE satisfies the identity

$$XQ = QQ_u.$$

What is the corresponding result for point symmetries of a system of PDEs with M dependent variables?

9.3 Calculate the travelling wave solutions of the nonlinear filtration equation (9.14). Use the remaining symmetries (9.15) to construct a family of solutions from the solutions that you have found.

9.4 The *Spherical Korteweg–de Vries equation,*

$$u_t + \frac{u}{t} + uu_x + u_{xxx} = 0,$$

has Lie point symmetries generated by

$$X_1 = \partial_x, \qquad X_2 = \ln t\,\partial_x + t^{-1}\partial_u,$$
$$X_3 = 3t\partial_t + x\partial_x - 2u\partial_u.$$

Find the most general solution that is invariant under the group generated by X_2. Now use the remaining Lie point symmetries to construct a two-parameter family of solutions from the invariant solution.

9.5 Calculate the ($\tau = 1$) nonclassical symmetries of Burgers' equation and find any invariant solutions.

10

Classification of Invariant Solutions

Instead of a million count half a dozen . . .
Simplify, simplify.

(H. D. Thoreau: Walden)

10.1 Equivalence of Invariant Solutions

Our aim in this chapter is to divide the set of all invariant solutions of a given differential equation into equivalence classes. Two invariant solutions are *equivalent* if one can be mapped to the other by a point symmetry of the PDE. Classification greatly simplifies the problem of determining all invariant solutions. It is only necessary to find one (general) invariant solution from each class; then the whole class can be constructed by applying the symmetries. This strategy minimizes the effort needed to obtain invariant solutions.

For simplicity, we restrict attention to the problem of equivalence of solutions that are invariant under a one-parameter Lie group of point symmetries. To avoid a proliferation of indices, let x and u denote the N independent and M dependent variables respectively, and let z be the set of all variables, that is, $z = (x, u)$. We shall examine what happens when a symmetry

$$\Gamma : z \mapsto \hat{z} \tag{10.1}$$

acts on a solution that is invariant under the one-parameter symmetry group generated by

$$X = \kappa^i X_i, \tag{10.2}$$

where each κ^i is a constant and the generators

$$X_i = \zeta_i^s(z)\partial_{z^s} \tag{10.3}$$

form a basis for the Lie algebra of point symmetry generators. Henceforth, we adopt the convention that a caret over a function or operator means that z has been replaced by $\hat{z}$. For instance,

$$\hat{X}_i = \zeta_i^s(\hat{z})\partial_{\hat{z}^s}; \tag{10.4}$$

here the functions ζ_i^s are exactly the same as in (10.3), although their argument has changed.

The procedure for generating families of invariant solutions should now be familiar (see §4.3 and Example 9.5). Suppose that the solution $u = f(x)$ is invariant under the one-parameter symmetry group generated by X. We write $u = f(x)$ in terms of $(\hat{x}, \hat{u})$ to obtain $\hat{u} = \tilde{f}(\hat{x})$. The symmetry group generator X is also written in terms of $(\hat{x}, \hat{u})$, and finally the carets are removed. This yields the generator of a one-parameter group under which the transformed solution $u = \tilde{f}(x)$ is invariant; we call this generator $\tilde{X}$.

The solutions $u = f(x)$ and $u = \tilde{f}(x)$ are equivalent, because a symmetry (Γ) maps one to the other. Similarly, the symmetry maps X to $\tilde{X}$, so these generators are regarded as equivalent. We aim to classify invariant solutions by classifying the associated symmetry generators. Having done this, one generator from each class is used to obtain the desired set of invariant solutions. A set consisting of exactly one generator from each class is called an *optimal system of generators*.

To classify the generators, we must write X in terms of $\hat{z}$. Recall from §2.6 that if $\hat{z} = e^{\varepsilon X_j} z$ then

$$e^{\varepsilon X_j} F(z) = F\left(e^{\varepsilon X_j} z\right) = F(\hat{z}) \tag{10.5}$$

for any smooth function F. More generally, we define the action of any symmetry (10.1) on any smooth function F similarly:

$$\Gamma F(z) = F(\Gamma z) = F(\hat{z}) \tag{10.6}$$

(This allows us to deal with discrete symmetries as well as Lie symmetries.) Now let $\Gamma(\delta)$ denote the one-parameter Lie group of symmetries generated by X. Here δ is the group parameter and

$$\Gamma(\delta) : z \mapsto e^{\delta X} z. \tag{10.7}$$

If F is an arbitrary smooth function then, from (10.6),

$$\hat{X} F(\hat{z}) = \Gamma X F(z) = \Gamma X \Gamma^{-1} F(\hat{z}).$$

Hence

$$\hat{X}^2 F(\hat{z}) = \Gamma X \Gamma^{-1} \Gamma X \Gamma^{-1} F(\hat{z}) = \Gamma X^2 \Gamma^{-1} F(\hat{z}),$$

and so on, and (assuming convergence) we can form the Lie series

$$\hat{\Gamma}(\delta) F(\hat{z}) = e^{\delta \hat{X}} F(\hat{z}) = \Gamma e^{\delta X} \Gamma^{-1} F(\hat{z}) = \Gamma \Gamma(\delta) \Gamma^{-1} F(\hat{z}).$$

As F is arbitrary, we conclude that

$$\hat{X} = \Gamma X \Gamma^{-1} \tag{10.8}$$

is the generator of the one-parameter Lie group of symmetries

$$\hat{\Gamma}(\delta) = \Gamma \Gamma(\delta) \Gamma^{-1}. \tag{10.9}$$

In Chapter 11, we shall use this result to obtain discrete symmetries. The remainder of the current chapter deals only with equivalence under Lie symmetries that are generated by a finite-dimensional Lie algebra with a basis $\{X_1, \ldots, X_R\}$. (We ignore classes of generators that depend on arbitrary functions, such as the infinite-dimensional subalgebras that occur in linear or linearizable PDEs.) It can be shown that this restricted equivalence problem is solvable by studying a finite sequence of one-dimensional problems. In each of these problems, we look at equivalence under the symmetries obtained from one of the generators in the basis,

$$\Gamma : z \mapsto \hat{z} = e^{\varepsilon X_j} z. \tag{10.10}$$

From (10.8),

$$\hat{X} = e^{\varepsilon X_j} X e^{-\varepsilon X_j} \tag{10.11}$$

for any generator X. (Here and for the rest of this chapter we do not sum over the index j, even when it is repeated, because X_j denotes a specific generator.) In particular (10.11) holds for $X = X_j$, which commutes with $e^{-\varepsilon X_j}$; therefore

$$\hat{X}_j = X_j. \tag{10.12}$$

We can now write any generator X in terms of $\hat{z}$ by solving (10.11) for X and using (10.12) to obtain

$$X = e^{-\varepsilon \hat{X}_j} \hat{X} e^{\varepsilon \hat{X}_j}. \tag{10.13}$$

For each ε, the right-hand side of (10.13) generates a one-parameter symmetry group under which $\hat{u} = \tilde{f}(\hat{x})$ is invariant. Therefore the transformed solution $u = \tilde{f}(x)$ is invariant under the group generated by

$$\tilde{X} = e^{-\varepsilon X_j} X e^{\varepsilon X_j}, \tag{10.14}$$

which is equivalent to X. This result holds for all Lie point symmetry generators X_j. Essentially, the classification problem for generators is solved by using (10.14), with various generators X_j in turn, to reduce every generator to the simplest equivalent form. We now discuss how to carry out this task.

10.2 How to Classify Symmetry Generators

The equivalence relation (10.14) involves symmetry generators, rather than solutions of any particular differential equation. Once we have classified the generators for a particular Lie algebra, the classification applies to every differential equation with that Lie algebra, whether it is an ODE or PDE. If we knew every possible Lie algebra, we could solve the classification problem once and for all. The problem of identifying all possible Lie algebras has been solved for scalar ODEs, but not for PDEs or systems of ODEs. Therefore we usually need to do the classification on a case-by-case basis.

From (10.14), $\tilde{X}$ satisfies the initial-value problem

$$\frac{d\tilde{X}}{d\varepsilon} = -X_j\tilde{X} + \tilde{X}X_j = -[X_j, \tilde{X}], \qquad \tilde{X}|_{\varepsilon=0} = X. \tag{10.15}$$

Differentiating the above ODE with respect to ε, we obtain

$$\frac{d^2\tilde{X}}{d\varepsilon^2} = -\left[X_j, \frac{d\tilde{X}}{d\varepsilon}\right] = (-1)^2[X_j, [X_j, \tilde{X}]],$$

and so on. Taylor's theorem leads to the following series solution, which is valid for all ε sufficiently close to zero:

$$\tilde{X} = X - \varepsilon[X_j, X] + \frac{\varepsilon^2}{2!}[X_j, [X_j, X]] - \cdots. \tag{10.16}$$

If X and X_j commute then (10.16) yields

$$\tilde{X} = X \qquad \forall\, \varepsilon, \tag{10.17}$$

so the generator X is unaltered. For abelian Lie algebras, all generators commute; thus no two linearly independent generators are equivalent. The "optimal system" of generators contains every generator!

For non-abelian Lie algebras, we aim to use each basis generator X_j to simplify X by eliminating as many of the constants κ^i as possible. One further simplification is possible: we can multiply X by any nonzero constant λ. (Recall that the group generated by λX is precisely the same as the group generated by X. Therefore the set of group-invariant solutions is unaffected when X is rescaled.)

Example 10.1 Consider the non-abelian two-dimensional Lie algebra with a basis $\{X_1, X_2\}$ such that

$$[X_1, X_2] = X_1.$$

[This Lie algebra is usually denoted by $\mathfrak{a}(1)$.] Each generator is of the form

$$X = \kappa^1 X_1 + \kappa^2 X_2.$$

We start by determining the set of generators that are equivalent to X under the group generated by X_1. From (10.16),

$$\begin{aligned}
\tilde{X} &= X - \varepsilon[X_1, X] + \frac{\varepsilon^2}{2!}[X_1, [X_1, X]] - \cdots \\
&= \kappa^i \left(X_i - \varepsilon[X_1, X_i] + \frac{\varepsilon^2}{2!}[X_1, [X_1, X_i]] - \cdots \right) \\
&= \kappa^1 X_1 + \kappa^2 (X_2 - \varepsilon X_1) \\
&= (\kappa^1 - \varepsilon \kappa^2) X_1 + \kappa^2 X_2.
\end{aligned}$$

In particular, if $\kappa^2 \neq 0$ then we can choose $\varepsilon = \kappa^1/\kappa^2$, to show that X is equivalent to $\kappa^2 X_2$. Rescaling, we set $\kappa^2 = 1$ without loss of generality (because of the assumption that $\kappa^2 \neq 0$). The only remaining possibility is that $\kappa^2 = 0$. Then $X = \kappa^1 X_1$ for some $\kappa^1 \neq 0$, and we rescale to set $\kappa^1 = 1$.

In summary, the set $\{X_1, X_2\}$ is an optimal system for $\mathfrak{a}(1)$. Every solution that is invariant under a one-parameter group generated by $X \in \mathfrak{a}(1)$ is equivalent to a solution that is invariant under the group generated by one of the generators in the optimal system.

In this very simple example, we have been able to solve the equivalence problem by using only the group generated by X_1. What would have happened if we had tried to use the group generated by X_2 instead? In that case,

$$\begin{aligned}
\tilde{X} &= \kappa^i \left(X_i - \varepsilon[X_2, X_i] + \frac{\varepsilon^2}{2!}[X_2, [X_2, X_i]] - \cdots \right) \\
&= e^{\varepsilon} \kappa^1 X_1 + \kappa^2 X_2.
\end{aligned}$$

This group acts on X by rescaling one of the components. If $\kappa^2 \neq 0$, we rescale X and set $\kappa^2 = 1$ without loss of generality. If $\kappa^1 \neq 0$ also, we may choose $\varepsilon = -\ln|\kappa^1|$ to get

$$\tilde{X} = \pm X_1 + X_2.$$

Although we have been able to simplify the generator considerably, we cannot reach the simplest equivalent form,

$$\tilde{X} = X_2,$$

without also using the group generated by X_1.

Usually, most or all of the one-parameter groups $e^{\varepsilon X_j}$ are needed to produce an optimal system of generators. The calculations are fairly simple for low-dimensional Lie algebras but increase in complexity as the dimension increases. Therefore, we now restate the equivalence problem in terms of matrices, in order to take advantage of computer algebra packages for matrix manipulation.

To begin with, we split X and $\tilde{X}$ into components as follows:

$$X = \kappa^i X_i, \tag{10.18}$$

$$\tilde{X} = \kappa^i \tilde{X}_i = \kappa^i e^{-\varepsilon X_j} X_i e^{\varepsilon X_j}. \tag{10.19}$$

Recall that any Lie algebra is closed under the commutator, so

$$\tilde{X}_i = \left(A(j,\varepsilon)\right)_i^m X_m \tag{10.20}$$

for some $R \times R$ matrix $A(j, \varepsilon)$. Therefore

$$\tilde{X} = \tilde{\kappa}^m X_m, \tag{10.21}$$

where

$$\tilde{\kappa}^m = \kappa^i \left(A(j,\varepsilon)\right)_i^m. \tag{10.22}$$

It is convenient to introduce the row vectors

$$\kappa = (\kappa^1, \ldots, \kappa^R),$$

$$\tilde{\kappa} = (\tilde{\kappa}^1, \ldots, \tilde{\kappa}^R).$$

Then Γ can be regarded as a map that acts on the constants κ as follows:

$$\Gamma : \kappa \mapsto \tilde{\kappa} = \kappa A(j, \varepsilon). \tag{10.23}$$

Once we know the matrix $A(j,\varepsilon)$ corresponding to each generator X_j, we try to solve the equivalence problem by applying these matrices to κ appropriately, each time choosing ε so as to simplify $\tilde{\kappa}$.

Our task now is to obtain $A(j,\varepsilon)$. The generator $\tilde{X}_i$ is the solution of the initial-value problem

$$\frac{d\tilde{X}_i}{d\varepsilon} = -e^{-\varepsilon X_j}[X_j, X_i]e^{\varepsilon X_j} = c_{ij}^k \tilde{X}_k,$$
$$\tilde{X}_i|_{\varepsilon=0} = X_i.$$

Therefore, from (10.20),

$$\frac{d\big(A(j,\varepsilon)\big)_i^m}{d\varepsilon} X_m = c_{ij}^k \big(A(j,\varepsilon)\big)_k^m X_m, \qquad \big(A(j,0)\big)_i^m X_m = X_i.$$

The generators X_m are linearly independent, so

$$\frac{d\big(A(j,\varepsilon)\big)_i^m}{d\varepsilon} = c_{ij}^k \big(A(j,\varepsilon)\big)_k^m, \qquad \big(A(j,0)\big)_i^m = \delta_i^m. \tag{10.24}$$

We now define the matrix $C(j)$, whose components are

$$\big(C(j)\big)_i^k = c_{ij}^k. \tag{10.25}$$

This enables us to write (10.24) as the matrix differential equation

$$\frac{dA(j,\varepsilon)}{d\varepsilon} = C(j)A(j,\varepsilon), \qquad A(j,0) = I, \tag{10.26}$$

whose general solution is

$$A(j,\varepsilon) = \exp\{\varepsilon C(j)\} = \sum_{n=0}^{\infty} C(j)^n \frac{\varepsilon^n}{n!}. \tag{10.27}$$

For low-dimensional Lie algebras, it is easy enough to calculate $A(j,\varepsilon)$ by hand, as the next example shows. Most computer algebra systems have linear algebra packages that will calculate the matrix exponential, $A(j,\varepsilon)$, given the matrix of structure constants, $C(j)$. Whether or not computer assistance is used, it is always wise to choose a basis in which as many structure constants as possible are zero to avoid making the task of simplifying κ unnecessarily difficult.

As we have seen, generators of abelian Lie algebras cannot be simplified at all, for each generator is invariant under the group generated by any other one.

Some non-abelian Lie algebras also have one or more invariants, $I(\kappa)$, such that

$$I\left(\kappa \exp[\varepsilon C(j)]\right) = I(\kappa), \qquad \forall\, j,\, \varepsilon. \tag{10.28}$$

Invariants act as constraints on the amount of simplification that is possible, so it is important to be able to derive them systematically. We now develop a technique for doing this.

Differentiating (10.28) with respect to ε at $\varepsilon = 0$ leads to the following necessary (and sufficient) condition for $I(\kappa)$ to be invariant:

$$\kappa C(j) \nabla I(\kappa) = 0, \qquad \forall\, j, \tag{10.29}$$

where

$$\nabla I(\kappa) = \begin{bmatrix} I_1(\kappa) \\ \vdots \\ I_R(\kappa) \end{bmatrix}, \qquad I_i(\kappa) \equiv \frac{\partial I(\kappa)}{\partial \kappa^i}.$$

The invariance conditions (10.29) can be solved by the method of characteristics. A convenient way to do this is to write (10.29) as

$$K(\kappa) \nabla I(\kappa) = \mathbf{0}, \tag{10.30}$$

where $K(\kappa)$ is the $R \times R$ matrix whose jth row is $\kappa C(j)$. The matrix PDE (10.30) may be simplified by reducing $K(\kappa)$ to echelon form; the reduced equations are usually easy to solve by the method of characteristics. If $\rho = \mathrm{Rank}(K(\kappa))$, there are $R - \rho$ functionally independent invariants.

Example 10.2 Here we use the matrix method to determine an optimal system of generators for the three-dimensional Lie algebra $\mathfrak{sl}(2)$. As usual, we choose a basis $\{X_1, X_2, X_3\}$ such that

$$[X_1, X_2] = X_1, \qquad [X_1, X_3] = 2X_2, \qquad [X_2, X_3] = X_3.$$

The only nonzero structure constants c_{ij}^k with $j = 1$ are

$$c_{21}^1 = -1, \qquad c_{31}^2 = -2,$$

so

$$C(1) = \begin{bmatrix} 0 & 0 & 0 \\ -1 & 0 & 0 \\ 0 & -2 & 0 \end{bmatrix}.$$

Similarly, the remaining nonzero structure constants give

$$C(2) = \begin{bmatrix} 1 & 0 & 0 \\ 0 & 0 & 0 \\ 0 & 0 & -1 \end{bmatrix}, \qquad C(3) = \begin{bmatrix} 0 & 2 & 0 \\ 0 & 0 & 1 \\ 0 & 0 & 0 \end{bmatrix}.$$

Now we exponentiate the matrices $\varepsilon C(j)$ to obtain

$$A(1,\varepsilon) = \begin{bmatrix} 1 & 0 & 0 \\ -\varepsilon & 1 & 0 \\ \varepsilon^2 & -2\varepsilon & 1 \end{bmatrix}, \qquad A(2,\varepsilon) = \begin{bmatrix} e^{\varepsilon} & 0 & 0 \\ 0 & 1 & 0 \\ 0 & 0 & e^{-\varepsilon} \end{bmatrix},$$

$$A(3,\varepsilon) = \begin{bmatrix} 1 & 2\varepsilon & \varepsilon^2 \\ 0 & 1 & \varepsilon \\ 0 & 0 & 1 \end{bmatrix}. \tag{10.31}$$

Before trying to simplify κ, let us first check for the existence of invariants. The row vectors $\kappa C(j)$ are

$$\kappa C(1) = (-\kappa^2, -2\kappa^3, 0),$$
$$\kappa C(2) = (\kappa^1, 0, -\kappa^3),$$
$$\kappa C(3) = (0, 2\kappa^1, \kappa^2),$$

so any invariants satisfy

$$\begin{bmatrix} -\kappa^2 & -2\kappa^3 & 0 \\ \kappa^1 & 0 & -\kappa^3 \\ 0 & 2\kappa^1 & \kappa^2 \end{bmatrix} \begin{bmatrix} I_1(\kappa) \\ I_2(\kappa) \\ I_3(\kappa) \end{bmatrix} = \begin{bmatrix} 0 \\ 0 \\ 0 \end{bmatrix}.$$

Here $\rho = 2$, so there is one invariant:

$$I = (\kappa^2)^2 - 4\kappa^1\kappa^3. \tag{10.32}$$

(The calculation leading to (10.32) is left as an exercise.) We cannot affect I except by rescaling X, which is equivalent to multiplying κ by a nonzero constant. As I is quadratic in the components of κ, rescaling can only multiply I by a positive constant. Therefore we must consider three distinct problems, namely $I > 0$, $I < 0$ and $I = 0$.

The vector κ is transformed by the matrices $A(j, \varepsilon)$ as follows:

$$\kappa A(1, \varepsilon) = (\kappa^1 - \varepsilon\kappa^2 + \varepsilon^2\kappa^3,\ \kappa^2 - 2\varepsilon\kappa^3,\ \kappa^3), \tag{10.33}$$

$$\kappa A(2, \varepsilon) = (e^{\varepsilon}\kappa^1,\ \kappa^2,\ e^{-\varepsilon}\kappa^3), \tag{10.34}$$

$$\kappa A(3, \varepsilon) = (\kappa^1,\ 2\varepsilon\kappa^1 + \kappa^2,\ \varepsilon^2\kappa^1 + \varepsilon\kappa^2 + \kappa^3). \tag{10.35}$$

Suppose that $I > 0$. Then (10.33) can be used to replace κ^1 by zero. [If $\kappa^3 \neq 0$, choose $\varepsilon = (\kappa^2 + \sqrt{I})/(2\kappa^3)$; otherwise (i.e., if $\kappa^3 = 0$), choose $\varepsilon = \kappa^1/\kappa^2$, bearing in mind that κ^2 and κ^3 cannot both be zero if I is positive.] Next we use (10.35) with $\varepsilon = -\kappa^3/\kappa^2$ to replace κ^3 by zero (if it is not already zero). Finally the generator is rescaled by setting $\kappa^2 = 1$. Therefore every generator with $I > 0$ is equivalent to X_2.

Now suppose that $I < 0$, which implies that $\kappa^1\kappa^3 > 0$. First replace κ^2 by zero, using (10.33) with $\varepsilon = \kappa^2/(2\kappa^3)$. Then use (10.34) with $\varepsilon = \frac{1}{2}\ln(\kappa^3/\kappa^1)$ to make κ^1 and κ^3 equal. After rescaling we find that every generator with $I < 0$ is equivalent to $X_1 + X_3$.

If $I = 0$ then either all three components of κ are nonzero or else κ^2 and one of κ^1, κ^3 are zero. We can reduce the first case to the second by using (10.35) with $\varepsilon = -\kappa^2/(2\kappa^1)$ to replace κ^2 and κ^3 by zero. Moreover

$$(0, 0, \kappa^3)A(1, 1)A(3, 1) = (\kappa^3, -2\kappa^3, \kappa^3)A(3, 1) = (\kappa^3, 0, 0),$$

so we can replace the second and third components of any κ by zero. Rescaling, we conclude that all generators for which $I = 0$ are equivalent to X_1.

Therefore $\{X_1,\ X_2,\ X_1 + X_3\}$ is an optimal system of generators for $\mathfrak{sl}(2)$.

In the last two examples, the number of generators in the optimal system coincides with the dimension (R) of the Lie algebra. This does not always happen; quite commonly, the number of inequivalent generators exceeds R.

10.3 Optimal Systems of Invariant Solutions

Having obtained an optimal system of generators, we can use the method described in §9.1 to calculate the associated invariant solutions. Any complete set of these solutions, from which all other invariant solutions can be derived, is called an *optimal system of invariant solutions*. There are two obstacles that we may encounter. The first is that a generator in the optimal system might not yield any invariant solutions.

Example 10.3 We shall attempt to construct an optimal set of invariant solutions of

$$y'' = y^{-3}, \tag{10.36}$$

whose Lie point symmetries are generated by

$$X_1 = \partial_x, \qquad X_2 = x\partial_x + \tfrac{1}{2}y\partial_y, \qquad X_3 = x^2\partial_x + xy\partial_y. \tag{10.37}$$

The Lie algebra is $\mathfrak{sl}(2)$ (with our usual structure constants). Example 10.2 shows that $\{X_1, X_2, X_1 + X_3\}$ is an optimal system of generators. No solution is invariant under the group generated by X_1, because the invariant curves are all of the form

$$y = c.$$

Neither does the group generated by X_2 yield any invariant solutions, for the invariant curves are

$$y = c\sqrt{x}.$$

Such a curve is a solution only if $c^4 = -4$, so there are no real-valued solutions. The group generated by $X_1 + X_3$ has two invariant solutions, namely

$$y = \pm\sqrt{1 + x^2}. \tag{10.38}$$

The action of the remaining symmetries on these two solutions yields the general solution of the ODE, which is

$$y = \pm\sqrt{c_1 + (x + c_2)^2/c_1}, \qquad c_1 > 0. \tag{10.39}$$

Recall that X_1 and X_2 are representatives of the classes of generators with $I = 0$ and $I > 0$, respectively. Might there be invariant solutions associated with other generators in these classes? In fact, there are no such solutions. (The derivation of this result is left as an exercise.)

The second obstacle is that the reduced equation(s) determining one or more invariant solutions may be too difficult to solve analytically. Even if we cannot obtain an optimal system, we may still be able to find some invariant solutions.

Example 10.4 Consider the linear PDE

$$u_t = u_{xx} - \frac{2}{x^2}u. \tag{10.40}$$

The finite-dimensional subalgebra $\mathcal{L}_0$ has a basis

$$\begin{aligned} X_1 &= \partial_t, \qquad X_2 = \tfrac{1}{2}x\partial_x + t\partial_t - \tfrac{1}{4}u\partial_u, \\ X_3 &= xt\partial_x + t^2\partial_t - \left(\tfrac{1}{4}x^2 + \tfrac{1}{2}t\right)u\partial_u, \qquad X_4 = u\partial_u, \end{aligned} \tag{10.41}$$

for which the only nonzero commutators are

$$[X_1, X_2] = X_1, \qquad [X_1, X_3] = 2X_2, \qquad [X_2, X_3] = X_3.$$

Thus the first three generators in the basis span an $\mathfrak{sl}(2)$ subalgebra (which is the derived subalgebra of $\mathcal{L}_0$) and X_4 commutes with all generators. There are two invariants $I(\kappa)$, namely

$$I^1 = (\kappa^2)^2 - 4\kappa^1\kappa^3, \qquad I^2 = \kappa^4.$$

The generators are classified almost exactly as in Example 10.2, but each generator has an arbitrary multiple of X_4 added to it. We must also consider the possibility that the first three components of κ are all zero; rescaling, we set $\kappa^4 = 1$ without loss of generality. Therefore the following system of generators is optimal:

$$X_1 + \mu X_4, \qquad X_2 + \mu X_4, \qquad X_1 + X_3 + \mu X_4, \qquad X_4.$$

(Here μ is an arbitrary constant.) To find an optimal system of solutions, we use each of these generators in turn. The solutions that are invariant under $X_1 + \mu X_4$ are of the form

$$u = e^{\mu t}F(x), \tag{10.42}$$

where

$$F'' - (\mu + 2x^{-2})F = 0.$$

The general solution of this ODE is

$$F(x) = \begin{cases} c_1(\sqrt{\mu} - x^{-1})e^{\sqrt{\mu}x} + c_2(\sqrt{\mu} + x^{-1})e^{-\sqrt{\mu}x}, & \mu \neq 0, \\ c_1x^2 + c_2x^{-1}, & \mu = 0, \end{cases} \tag{10.43}$$

and so the general invariant solution is obtained by substituting (10.43) into (10.42). Similarly, the most general solution that is invariant under $X_2 + \mu X_4$ is

$$u = t^{\mu - 1/4}F(r), \qquad r = xt^{-1/2}, \tag{10.44}$$

where

$$F'' + \tfrac{1}{2}rF' + \left(\tfrac{1}{4} - \mu - 2r^{-2}\right)F = 0. \tag{10.45}$$

The solution of (10.45) can be expressed in terms of confluent hypergeometric functions; it is rather messy! A simpler form is possible for some values of μ; for example, when $\mu = 1/4$, one solution of the form (10.44) is

$$u = x^{-1} t^{1/2} e^{-\frac{x^2}{4t}}.$$

The solutions that are invariant under $X_1 + X_3 + \mu X_4$ are of the form

$$u = (1+t^2)^{-1/4} \exp\left\{\mu \tan^{-1} t - \frac{x^2 t}{4(1+t^2)}\right\} F(r), \qquad r = x(1+t^2)^{-1/2}, \tag{10.46}$$

where

$$F'' + \left(\tfrac{1}{4}r^2 - \mu - 2r^{-2}\right) F = 0. \tag{10.47}$$

These solutions are also combinations of confluent hypergeometric functions and elementary functions. Finally, the group generated by X_4 leaves only the trivial solution $u = 0$ invariant.

After an optimal system of invariant solutions is found, it is often quite easy to generate each class of invariant solutions, as described in §9.2. Although most PDEs have many solutions that are not invariant under any point symmetry, the invariant solutions typically describe limiting behaviour (especially for solutions of parabolic PDEs). Similarity (scaling-invariant) solutions are particularly useful in this respect.

Notes and Further Reading

PDEs with $N \geq 3$ independent variables can be reduced to ODEs if one looks for solutions that are invariant under an $(N-1)$-parameter Lie (sub)group of symmetries. It is possible to construct an optimal set of such reductions, building on the results for one-parameter groups (see Ovsiannikov (1982) for some further details).

Exercises

10.1 Show that (10.32) is the only invariant of $\mathfrak{sl}(2)$.

10.2 Calculate an optimal system of generators for $\mathfrak{so}(3)$, in a basis whose nonzero structure constants are (5.33). Are there any invariants $I(\kappa)$?

10.3 Consider the equivalence problem for the symmetries generated by the six-dimensional subalgebra of the heat equation [excluding generators of the form $X = U(x,t)\partial_u$]. These generators may be found at the end of the book, in the solution to Exercise 8.3. Calculate two functionally independent invariants, and hence derive an optimal system of generators.

10.4 Show that κ^i is invariant if $X_i \notin [\mathcal{L}, \mathcal{L}]$. Is this condition necessary to ensure that κ^i is invariant?

10.5 Show that the ODE in Example 10.3 has no real-valued solutions that are invariant under any group whose generator has $I > 0$ (in the notation of Example 10.2).

10.6 Calculate an optimal system of generators for the free convection problem described in Example 9.4. Use your results to obtain an optimal system of invariant solutions. (N.B. This system is very large, but most solutions are easy to obtain.)

11

Discrete Symmetries

There is a place apart
Beyond the solar ray,
Where parallel straight lines can meet
in an unofficial way.

(G. K. Chesterton: The Higher Mathematics)

11.1 Some Uses of Discrete Symmetries

In Chapter 1, the discrete symmetries of a triangle are used to introduce the reader to Lie symmetries. It seems appropriate that this closing chapter introduces a method that uses Lie symmetries to reveal the discrete symmetries of a given differential equation. Here are some reasons why it is important to determine the discrete symmetries.

(i) Discrete point symmetries are used to increase the efficiency of computational methods. If a boundary-value problem (BVP) is symmetric and the solution is known to be unique, computation can be carried out on a reduced domain. Alternatively, a spectral method can be used, with basis functions that are invariant under the symmetry. The presence of a discrete symmetry also improves the accuracy of some numerical methods.

(ii) Many nonlinear BVPs have multiple solutions, and it is necessary to identify when and how the system changes its behaviour as any parameters vary. Discrete symmetries must be taken into account, because the behaviour of "generic" nonsymmetric systems is usually quite different from that of systems with symmetries. It is important to identify all of the symmetries in a problem in order to understand its behaviour correctly.

(iii) Like Lie symmetries, discrete symmetries can be used to generate new solutions from known solutions. They can also be used to simplify an optimal system of generators. If two generators in the optimal system are related by a discrete symmetry, only one is needed; the other can be removed.

(iv) Discrete symmetries involving charge conjugation, parity change, and time reversal are central in quantum field theories. (Other discrete symmetries are important in physics, but we do not discuss them here.)

(v) Although we shall restrict our attention to discrete point symmetries, other types of discrete symmetries are also useful. The Legendre transformation is perhaps the best-known example of a discrete contact transformation. Such transformations can occur as symmetries of differential equations, even if there are no Lie contact symmetries. Auto-Bäcklund transformations are nonpoint discrete symmetries; they enable the user to construct hierarchies of solutions to nonlinear integrable PDEs.

The symmetry condition for discrete symmetries is almost always too hard to solve by a direct approach. This is because the determining equations typically form a highly coupled nonlinear system. Sometimes it is possible to simplify this system with the aid of computer algebra. An ansatz may lead to some discrete symmetries; however, there is no guarantee of obtaining all of them.

Nearly all problems that arise from applications have at least a one-parameter Lie group of point symmetries. If a given differential equation has a known finite-dimensional Lie algebra of symmetry generators, the direct approach described above is no longer necessary. Discrete and Lie symmetries interact in a way that can be used to derive the discrete symmetries systematically. (The same ideas also work for infinite-dimensional Lie algebras, but extra care is needed; we shall not discuss these algebras further.)

11.2 How to Obtain Discrete Symmetries from Lie Symmetries

From here on we use the notation introduced in §10.1, which readers may wish to review before continuing. Let

$$\Gamma : z \mapsto \hat{z} \tag{11.1}$$

be an arbitrary symmetry of a given differential equation, where z denotes the M dependent and N independent variables. We assume that the Lie algebra $\mathcal{L}$ of Lie point symmetry generators is R dimensional, and that the generators

$$X_i = \zeta_i^s(z)\partial_{z^s}, \qquad i = 1, \ldots, R, \tag{11.2}$$

form a basis for $\mathcal{L}$. In §10.1 we showed that if $X \in \mathcal{L}$, then

$$\hat{X} = \Gamma X \Gamma^{-1}$$

generates a one-parameter Lie group of point symmetries of the differential equation. The Lie algebra $\mathcal{L}$ is the set of all generators of Lie point symmetries. Consequently $\hat{X} \in \mathcal{L}$. In particular, each basis generator

$$\hat{X}_i = \Gamma X_i \Gamma^{-1} = \zeta_i^s(\hat{z}) \partial_{\hat{z}^s} \tag{11.3}$$

is in $\mathcal{L}$. Furthermore, the set of generators $\{\hat{X}_1, \ldots, \hat{X}_R\}$ is a basis for $\mathcal{L}$, because it is simply the original basis with $\hat{z}$ replacing z. Therefore each X_i can be written as a linear combination of the $\hat{X}_i$'s, as follows:

$$X_i = b_i^l \hat{X}_l. \tag{11.4}$$

The coefficients b_i^l are constants which are determined by the symmetry Γ and the basis $\{X_1, \ldots, X_R\}$. It is useful to regard these coefficients as elements of an $R \times R$ matrix

$$B = \left(b_i^l\right). \tag{11.5}$$

The linear equations (11.4) constitute a transformation between two bases (namely, the X_i's and the $\hat{X}_i$'s). Therefore the matrix B is nonsingular.

We use (11.4) to construct discrete symmetries as follows. First apply (11.4) to each of the variables $\hat{z}^s$ in turn, to obtain

$$\zeta_i^r(z) \frac{\partial \hat{z}^s}{\partial z^r} = b_i^l \hat{X}_l \hat{z}^s = b_i^l \zeta_l^s(\hat{z}), \qquad 1 \le i \le R, \quad 1 \le s \le M + N. \tag{11.6}$$

This system of $(M + N)R$ partial differential equations can be solved by the method of characteristics, which yields $\hat{z}$ in terms of z, the unknown constants b_i^l, and some arbitrary functions or constants of integration. By construction, every symmetry (discrete or otherwise) satisfies (11.6) for some matrix B, although (11.6) may also have solutions that are not symmetries. However, it is usually easy for us to identify all of the symmetries, by substituting the general solution of (11.6) into the symmetry condition.

As we already know the Lie point symmetries, we can discard them (i.e., factor them out) at any convenient stage in the calculation. This will leave us with a list of *inequivalent* discrete point symmetries that cannot be mapped to one another by any Lie point symmetry.

Example 11.1 To demonstrate the method in a simple context, we now determine the discrete symmetries of the ODE

$$y'' = \tan y', \tag{11.7}$$

whose Lie algebra of point symmetry generators has a basis

$$X_1 = \partial_x, \qquad X_2 = \partial_y. \tag{11.8}$$

Here $z = (x, y)$ and therefore (11.6) is

$$\begin{bmatrix} \hat{x}_x & \hat{y}_x \\ \hat{x}_y & \hat{y}_y \end{bmatrix} = \begin{bmatrix} b_1^1 & b_1^2 \\ b_2^1 & b_2^2 \end{bmatrix} \begin{bmatrix} 1 & 0 \\ 0 & 1 \end{bmatrix}.$$

The general solution of this system is

$$(\hat{x}, \hat{y}) = \left(b_1^1 x + b_2^1 y + c_1,\ b_1^2 x + b_2^2 y + c_2\right). \tag{11.9}$$

At this stage, it is worth simplifying (11.9) by factoring out the Lie symmetries. The translations generated by $\hat{X}_1$ and $\hat{X}_2$ add arbitrary constants to $\hat{x}$ and $\hat{y}$ respectively. We need only study the solution

$$(\hat{x}, \hat{y}) = \left(b_1^1 x + b_2^1 y,\ b_1^2 x + b_2^2 y\right), \tag{11.10}$$

because the remaining solutions can be generated from this one by using the Lie symmetries. Every discrete symmetry is of the form (11.10) for some matrix B (up to equivalence under translations). To find out which matrices correspond to discrete symmetries, we substitute (11.10) into the symmetry condition

$$\hat{y}'' = \tan \hat{y}' \qquad \text{when} \quad y'' = \tan y'. \tag{11.11}$$

The prolongation formulae give

$$\hat{y}' = \frac{b_1^2 + b_2^2 y'}{b_1^1 + b_2^1 y'},$$

$$\hat{y}'' = \frac{J y''}{\left(b_1^1 + b_2^1 y'\right)^3}, \qquad \text{where} \quad J \equiv \det(B) \neq 0.$$

Therefore the symmetry condition is

$$\frac{J \tan y'}{\left(b_1^1 + b_2^1 y'\right)^3} = \tan\left(\frac{b_1^2 + b_2^2 y'}{b_1^1 + b_2^1 y'}\right). \tag{11.12}$$

Differentiating (11.12) with respect to y', we obtain

$$\begin{aligned} \frac{J(1 + \tan^2 y')}{\left(b_1^1 + b_2^1 y'\right)^3} - \frac{3 b_2^1 J \tan y'}{\left(b_1^1 + b_2^1 y'\right)^4} &= \frac{J}{\left(b_1^1 + b_2^1 y'\right)^2}\left\{1 + \tan^2\left(\frac{b_1^2 + b_2^2 y'}{b_1^1 + b_2^1 y'}\right)\right\} \\ &= \frac{J}{\left(b_1^1 + b_2^1 y'\right)^2}\left\{1 + \frac{J^2 \tan^2 y'}{\left(b_1^1 + b_2^1 y'\right)^6}\right\}. \end{aligned} \tag{11.13}$$

If b_2^1 were nonzero, (11.13) would be an algebraic equation for $\tan y'$ in terms of y'. This cannot be so, for tan is a transcendental function. Thus $b_2^1 = 0$, which leads to $b_1^1 = 1$, $b_2^2 = \alpha \in \{-1, 1\}$; hence (11.12) is reduced to

$$\alpha \tan y' = \tan\left(\alpha y' + b_1^2\right).$$

Therefore the inequivalent discrete symmetries are

$$(\hat{x}, \hat{y}) = (x, \ \alpha y + q\pi x), \qquad \alpha \in \{-1, 1\}, \quad q \in \mathbb{Z}. \tag{11.14}$$

This example illustrates the basic method. The calculations are usually fairly easy if $\mathcal{L}$ is low dimensional. (To make them as simple as possible, work in coordinates that are canonical for one generator.) The number of unknown coefficients b_i^l increases rapidly with R; if $\mathcal{L}$ is abelian and $R > 2$, computer algebra should be used. If $\mathcal{L}$ is non-abelian, however, it is possible to factor out Lie symmetries *before* solving equations (11.6). Typically, this reduces the number of nonzero coefficients in B from R^2 to R. Then it is possible to find the discrete symmetries of differential equations for which R is not small, with little more effort than is needed to determine the Lie symmetries. Essentially, B is simplified by the same method that is used in §10.2 to classify the generators of one-parameter symmetry groups. We now investigate this in detail in order to be able to deal with differential equations for which $R > 2$.

11.3 Classification of Discrete Symmetries

If $\mathcal{L}$ is non-abelian then at least some of the commutators

$$[X_i, X_j] = c_{ij}^k X_k \tag{11.15}$$

are nonzero. Therefore the generators belong to equivalence classes (with more than one element), which we shall use to simplify B. Recall that X_i is equivalent to

$$\tilde{X}_i = \left(A(j, \varepsilon)\right)_i^p X_p$$

under the Lie symmetries generated by X_j. We can rewrite (11.4) as

$$\tilde{X}_i = \tilde{b}_i^l \hat{X}_l, \tag{11.16}$$

where

$$\tilde{b}_i^l = \left(A(j, \varepsilon)\right)_i^p b_p^l.$$

Therefore (11.16) is equivalent to

$$X_i = \tilde{b}_i^l \hat{X}_l. \tag{11.17}$$

Consequently, the solutions $\hat{z}$ of (11.6) are related to the solutions of

$$\zeta_i^r(z)\frac{\partial \hat{z}^s}{\partial z^r} = \tilde{b}_i^l \zeta_l^s(\hat{z}), \tag{11.18}$$

by symmetries in the one-parameter group generated by X_j. In other words, B is equivalent to the family of matrices $A(j,\varepsilon)B$. To factor out the Lie symmetries generated by X_j, we solve (11.18) for just one (simple) matrix in this family. Often, we are able to replace one or more elements of B by zero.

Similarly, the generator $\hat{X}_l$ is equivalent to $(A(j,\varepsilon))_l^p \hat{X}_p$ under the Lie symmetries generated by $\hat{X}_j$. Therefore, by the same argument as above, B is equivalent to $BA(j,\varepsilon)$. Henceforth the term *equivalence transformation* is used to describe the replacement of B by either $BA(j,\varepsilon)$ or $A(j,\varepsilon)B$.

For abelian Lie algebras, the elements of B are unrelated. This is not so for non-abelian Lie algebras, and we now derive the relationships between the elements. These relationships, together with the equivalence transformations, usually enable us to reduce B to a very simple form.

The symmetry generators $\hat{X}_i$ satisfy the same commutator relations as X_i, because each $\hat{X}_i$ is obtained from the corresponding X_i merely by replacing z with $\hat{z}$. For example, the generators

$$X_1 = \partial_x, \qquad X_2 = x\partial_x$$

have the commutator

$$[X_1, X_2] = [\partial_x, x\partial_x] = \partial_x = X_1.$$

If $\hat{X}_1$ and $\hat{X}_2$ are defined similarly, but with $\hat{x}$ replacing x, then

$$[\hat{X}_1, \hat{X}_2] = [\partial_{\hat{x}}, \hat{x}\partial_{\hat{x}}] = \partial_{\hat{x}} = \hat{X}_1.$$

Therefore the structure constants are unaltered by the change of basis. The same is true in general: if the generators X_i satisfy (11.15), then

$$[\hat{X}_i, \hat{X}_j] = c_{ij}^k \hat{X}_k \tag{11.19}$$

(with the same structure constants). Now we substitute $X_i = b_i^l \hat{X}_l$ into (11.15) to obtain

$$b_i^l b_j^m [\hat{X}_l, \hat{X}_m] = c_{ij}^k b_k^n \hat{X}_n.$$

Then (11.19) leads to the useful identities

$$c^n_{lm} b^l_i b^m_j = c^k_{ij} b^n_k. \tag{11.20}$$

These equations are nonlinear constraints on the elements of B. The constraints with $i \geq j$ are essentially the same as those with $i < j$ (the proof of this is left as an exercise). Therefore we restrict attention to the constraints (11.20) for which $i < j$. The constraints are not affected by any equivalence transformation. (Again, the proof is left as an exercise.) Furthermore, the order in which the matrices $A(j, \varepsilon)$ are used does not affect the classification of the matrices B; any ordering gives the same final form, provided that the parameters ε are chosen appropriately. Therefore we may use the constraints and the equivalence transformations in whichever order is most convenient.

Example 11.2 Consider the two-dimensional non-abelian Lie algebra $\mathfrak{a}(1)$, with a basis $\{X_1, X_2\}$ such that $[X_1, X_2] = X_1$. The only nonzero structure constants are

$$c^1_{12} = 1, \qquad c^1_{21} = -1. \tag{11.21}$$

The constraints (11.20) (with $i < j$) are

$$\begin{aligned} b^1_1 b^2_2 - b^2_1 b^1_2 &= b^1_1, \qquad && (i, j, n) = (1, 2, 1), \\ 0 &= b^2_1, && (i, j, n) = (1, 2, 2), \end{aligned}$$

and therefore (bearing in mind that B is nonsingular)

$$B = \begin{bmatrix} b^1_1 & 0 \\ b^1_2 & 1 \end{bmatrix}, \qquad b^1_1 \neq 0. \tag{11.22}$$

We now try to simplify B further by using equivalence transformations. The matrices $A(j, \varepsilon) = \exp\{\varepsilon C(j)\}$ are

$$A(1, \varepsilon) = \begin{bmatrix} 1 & 0 \\ -\varepsilon & 1 \end{bmatrix}, \qquad A(2, \varepsilon) = \begin{bmatrix} e^{\varepsilon} & 0 \\ 0 & 1 \end{bmatrix}. \tag{11.23}$$

Postmultiplying B by $A(1, \varepsilon)$, we obtain

$$BA(1, \varepsilon) = \begin{bmatrix} b^1_1 & 0 \\ b^1_2 - \varepsilon & 1 \end{bmatrix}.$$

Therefore, by choosing $\varepsilon = b_2^1$, we can replace b_2^1 by zero. Then

$$BA(2,\varepsilon) = \begin{bmatrix} e^{\varepsilon} b_1^1 & 0 \\ 0 & 1 \end{bmatrix},$$

which can be simplified by setting $\varepsilon = -\ln |b_1^1|$; this is equivalent to replacing b_1^1 by ± 1. No further simplification is possible, so we are left with two inequivalent matrices:

$$B = \begin{bmatrix} \alpha & 0 \\ 0 & 1 \end{bmatrix}, \qquad \text{where} \quad \alpha \in \{-1, 1\}. \tag{11.24}$$

N.B. Exactly the same result can be obtained by premultiplying (11.22) by $A(j, \varepsilon)$, although different choices of ε are needed.

Example 11.3 In this example, we identify the inequivalent matrices B associated with $\mathfrak{sl}(2)$. We shall work in our usual basis, for which the only nonzero structure constants are

$$c_{12}^1 = -c_{21}^1 = 1, \qquad c_{13}^2 = -c_{31}^2 = 2, \qquad c_{23}^3 = -c_{32}^3 = 1. \tag{11.25}$$

From (10.31),

$$A(1,\varepsilon) = \begin{bmatrix} 1 & 0 & 0 \\ -\varepsilon & 1 & 0 \\ \varepsilon^2 & -2\varepsilon & 1 \end{bmatrix}, \qquad A(2,\varepsilon) = \begin{bmatrix} e^{\varepsilon} & 0 & 0 \\ 0 & 1 & 0 \\ 0 & 0 & e^{-\varepsilon} \end{bmatrix},$$
$$A(3,\varepsilon) = \begin{bmatrix} 1 & 2\varepsilon & \varepsilon^2 \\ 0 & 1 & \varepsilon \\ 0 & 0 & 1 \end{bmatrix}. \tag{11.26}$$

The constraints (11.20) are highly coupled (because $\mathfrak{sl}(2)$ is a simple Lie algebra). For example, the constraints with $n = 1$ are

$$b_1^1 b_2^2 - b_1^2 b_2^1 = b_1^1, \tag{11.27}$$

$$b_1^1 b_3^2 - b_1^2 b_3^1 = 2b_2^1, \tag{11.28}$$

$$b_2^1 b_3^2 - b_2^2 b_3^1 = b_3^1. \tag{11.29}$$

(The other constraints are left for the reader to find and use.)

If $b_1^1 \neq 0$, we premultiply B by $A(1,\ b_2^1/b_1^1)$, which is equivalent to setting $b_2^1 = 0$. Then (11.27) gives $b_2^2 = 1$, and so (11.29) is satisfied if $b_3^1 = 0$. Consequently (11.28) yields $b_3^2 = 0$. So far, we have reduced B to the form

$$B = \begin{bmatrix} b_1^1 & b_1^2 & b_1^3 \\ 0 & 1 & b_2^3 \\ 0 & 0 & b_3^3 \end{bmatrix}. \tag{11.30}$$

Next we postmultiply (11.30) by $A(3,\ -b_1^2/(2b_1^1))$ to set $b_1^2 = 0$. Then it turns out that the remaining constraints (11.20) are satisfied only if $b_1^3 = b_2^3 = 0$ and $b_3^3 = 1/b_1^1$. Finally, we premultiply B by $A(2,\ -\ln|b_1^1|)$ to obtain two inequivalent matrices:

$$B = \begin{bmatrix} \alpha & 0 & 0 \\ 0 & 1 & 0 \\ 0 & 0 & \alpha \end{bmatrix}, \qquad \alpha \in \{-1, 1\}. \tag{11.31}$$

The only remaining possiblity is that $b_1^1 = 0$. Applying a similar procedure to the above, we find that

$$B = \begin{bmatrix} 0 & 0 & \alpha \\ 0 & -1 & 0 \\ \alpha & 0 & 0 \end{bmatrix}, \qquad \alpha \in \{-1, 1\}. \tag{11.32}$$

So there are four distinct matrices B that are inequivalent under Lie symmetries.

11.4 Examples

All of the results in the previous section depend only on the given Lie algebra (not on any particular differential equation). The advantage of such general results is that they can be applied directly to any differential equation having that Lie algebra. On one hand, each inequivalent B may lead to more than one discrete symmetry of a particular differential equation. On the other hand, some matrices B that satisfy the constraints (11.20) may not correspond to even one discrete symmetry. We end this chapter with three examples to show how the method works in practice.

Example 11.4 Consider the ODE

$$y''' = \frac{y''^2}{x} - \frac{y''}{y'} \tag{11.33}$$

whose Lie algebra of point symmetry generators has a basis

$$X_1 = \partial_y, \qquad X_2 = \tfrac{1}{2}x\partial_x + y\partial_y. \tag{11.34}$$

In this basis, $[X_1, X_2] = X_1$, so (from Example 11.2) the inequivalent discrete symmetries satisfy

$$\begin{bmatrix} X_1\hat{x} & X_1\hat{y} \\ X_2\hat{x} & X_2\hat{y} \end{bmatrix} = \begin{bmatrix} \alpha & 0 \\ 0 & 1 \end{bmatrix} \begin{bmatrix} 0 & 1 \\ \frac{1}{2}\hat{x} & \hat{y} \end{bmatrix} = \begin{bmatrix} 0 & \alpha \\ \frac{1}{2}\hat{x} & \hat{y} \end{bmatrix},$$

whose general solution is

$$\hat{x} = c_1 x, \qquad \hat{y} = \alpha y + c_2 x^2, \qquad \alpha \in \{-1, 1\}. \tag{11.35}$$

The symmetry condition is satisfied if and only if $\alpha c_1^2 = 1$ and $c_2 = 0$. Therefore the only real-valued discrete point symmetries (up to equivalence) are

$$(\hat{x}, \hat{y}) \in \{(x, y), (-x, y)\}. \tag{11.36}$$

Both of these correspond to $\alpha = 1$. There are no real-valued symmetries with $\alpha = -1$ (although there are two inequivalent complex-valued symmetries).

Example 11.5 The *Chazy equation*

$$y''' = 2yy'' - 3y'^2 + \lambda(6y' - y^2)^2, \tag{11.37}$$

has the following basis for its Lie algebra:

$$X_1 = \partial_x, \qquad X_2 = x\partial_x - y\partial_y, \qquad X_3 = x^2\partial_x - (2xy + 6)\partial_y. \tag{11.38}$$

The commutators of these generators are

$$[X_1, X_2] = X_1, \qquad [X_1, X_3] = 2X_2, \qquad [X_2, X_3] = X_3,$$

and therefore the Lie algebra is $\mathfrak{sl}(2)$.

For the Chazy equation, the system (11.6) is

$$\begin{bmatrix} X_1\hat{x} & X_1\hat{y} \\ X_2\hat{x} & X_2\hat{y} \\ X_3\hat{x} & X_3\hat{y} \end{bmatrix} = B \begin{bmatrix} 1 & 0 \\ \hat{x} & -\hat{y} \\ \hat{x}^2 & -(2\hat{x}\hat{y} + 6) \end{bmatrix}, \tag{11.39}$$

where B is one of the four matrices (11.31), (11.32). It turns out that there are two solutions of (11.39) for each B. For example, if B is the identity matrix then

$$(\hat{x}, \hat{y}) \in \{(x, y), (x + 6/y, -y)\}.$$

However, only the first of these is a symmetry of the Chazy equation; the other solution violates the symmetry condition. Each matrix B generates precisely one discrete symmetry of the Chazy equation; the complete list is

$$(\hat{x}, \hat{y}) \in \left\{(x, y), (-x, -y), (-1/x, x^2 y + 6x), (1/x, -(x^2 y + 6x))\right\}. \tag{11.40}$$

Example 11.6 The above method works for PDEs and ODEs, scalar equations and systems. Consider the *Harry–Dym equation*,

$$u_t = u^3 u_{xxx}, \tag{11.41}$$

which has a five-dimensional Lie algebra of point symmetry generators. The basis

$$X_1 = \partial_x, \qquad X_2 = x\partial_x + u\partial_u, \qquad X_3 = x^2\partial_x + 2xu\partial_u,$$
$$X_4 = \partial_t, \qquad X_5 = t\partial_t - \tfrac{1}{3}u\partial_u$$

has the following nonzero structure constants c_{ij}^k, $i < j$:

$$c_{12}^1 = 1, \qquad c_{13}^2 = 2, \qquad c_{23}^3 = 1, \qquad c_{45}^4 = 1.$$

Note that the first three generators form a basis for an $\mathfrak{sl}(2)$ subalgebra, and the other two generators span an $\mathfrak{a}(1)$ subalgebra. It turns out that the inequivalent matrices B incorporate the $\mathfrak{sl}(2)$ and $\mathfrak{a}(1)$ matrices as subblocks: either

$$(X_1, X_2, X_3, X_4, X_5) = (\alpha\hat{X}_1, \hat{X}_2, \alpha\hat{X}_3, \beta\hat{X}_4, \hat{X}_5), \tag{11.42}$$

or

$$(X_1, X_2, X_3, X_4, X_5) = (\alpha\hat{X}_3, -\hat{X}_2, \alpha\hat{X}_1, \beta\hat{X}_4, \hat{X}_5), \tag{11.43}$$

where α, β are each either 1 or -1.

The general solution of (11.6) with (11.42) is

$$(\hat{x}, \hat{t}, \hat{u}) = (\alpha x, \beta t, cu).$$

Substituting this result into the symmetry condition, we obtain $c = \alpha\beta$. The discrete symmetries corresponding to (11.43) are found similarly. In all, there are eight inequivalent real discrete symmetries:

$$(\hat{x}, \hat{t}, \hat{u}) \in \left\{ (\alpha x, \beta t, \alpha\beta u), \left(-\frac{\alpha}{x}, \beta t, \frac{\alpha\beta u}{x^2}\right) \right\}, \qquad \alpha, \beta \in \{-1, 1\}. \tag{11.44}$$

Further Reading

Bifurcation theory describes the way in which nonlinear systems change their behaviour as parameters are varied. For systems with symmetries, *equivariant bifurcation theory* is needed to deal with degeneracies that are associated with the symmetries. Golubitsky, Stewart, and Schaeffer (1988) is a comprehensive introduction to this fascinating subject.

In §11.3, we studied a set of linear transformations of the Lie algebra. These transformations, which are represented by the matrix B, are called *automorphisms*. For more information about Lie algebras in general, and automorphisms in particular, I recommend Fuchs and Schweigert (1997) (which is aimed at physicists).

The Chazy equation has many interesting properties, as described in Clarkson and Olver (1996). This paper develops the methods outlined in §6.3 and applies them to the Chazy equation.

The methods described in this chapter are fairly new. To learn more, consult Hydon (1998a,b).

Exercises

11.1 Find a set of inequivalent real-valued discrete symmetries of

$$y'' = \frac{y'}{x} + \frac{4y^2}{x^3},$$

whose Lie symmetries are generated by $X_1 = x\partial_x + y\partial_y$.

11.2 Show that the constraints (11.20) with $i < j$ are sufficient to describe the whole set (i.e., the constraints with $i \geq j$ give nothing new).

11.3 Show that the constraints (11.20) are unchanged if B is replaced by either $A(j, \varepsilon)B$ or $BA(j, \varepsilon)$. Hint: First show that

$$c^n_{lm}\left(A(j,\varepsilon)\right)^l_i\left(A(j,\varepsilon)\right)^m_j = c^k_{ij}\left(A(j,\varepsilon)\right)^n_k.$$

11.4 Derive the results obtained in Example 11.3, by writing out the (nine) constraints (11.20) in detail.

11.5 Find a set of inequivalent matrices B for the Lie algebra $\mathfrak{so}(3)$.

11.6 Calculate a set of inequivalent discrete symmetries of the ODE (5.9).

11.7 Calculate a set of inequivalent discrete symmetries of Burgers' equation.

11.8 Calculate a set of inequivalent discrete symmetries of the nonlinear filtration equation (9.26). This equation is invariant under the hodograph transformation $(x, t, u) \mapsto (u, t, x)$; if this transformation is not in your set of inequivalent discrete symmetries, show how it can be derived from a member of the set.

11.9 Use the results of Exercise 3.6 to calculate a set of inequivalent discrete symmetries of

$$y'' = (1 - y')^3.$$

11.10 To find discrete contact symmetries, the generators X_i in (11.6) are prolonged once (because $\hat{z}$ depends upon x, u, and first derivatives of u). The only Lie contact symmetries of the ODE (11.33) are the point symmetries (11.34). Show that this ODE has nonpoint discrete contact symmetries. (N.B. Remember that for a contact transformation, $\hat{x}$, $\hat{y}$, and $\hat{y}'$ depend on x, y, and y' only.)

Hints and Partial Solutions to Some Exercises

Chapter 1

1.1 The general solution is $y = cx$, so the solutions are straight lines passing through the origin. The most obvious symmetries are reflections, rotations, and scalings of the form $(\hat{x}, \hat{y}) = (kx, ky)$ (where k is a positive constant).

1.2 The general solution of the ODE is

$$y = \frac{cx^2 - 1}{cx^2 + 1};$$

the symmetries are scalings in the x direction.

1.4 $\alpha = 2$.

1.5 The linear superposition principle: if $y = y(x)$ is any solution of the inhomogeneous ODE and $y = y_0(x)$ is any solution of the homogeneous ODE $y' = F(x)y$, then $\hat{y} = y(x) + y_0(x)$ is a solution of the inhomogeneous ODE. The general solution of the homogeneous ODE is

$$y = \varepsilon \exp\left\{ \int F(x)\, dx \right\}.$$

Chapter 2

2.1 (a) $X = \partial_x + \partial_y, \quad (r, s) = (y - x,\ x),$

(b) $X = xy\partial_x + y^2\partial_y, \quad (r, s) = (y/x,\ -1/y).$

2.2 (a) $(\hat{x}, \hat{y}) = (x + \varepsilon,\ e^{\varepsilon} y),$

(b) $(\hat{x}, \hat{y}) = \left(\frac{x\cos\varepsilon + \sin\varepsilon}{\cos\varepsilon - x\sin\varepsilon},\ \frac{y}{\cos\varepsilon - x\sin\varepsilon}\right),$

(c) Hint: use canonical coordinates.

2.3 If $\alpha \neq 0$, the origin $(0, 0)$ is the only invariant point. (If $\alpha = 0$, every point on the line $x = 0$ is invariant.) The symmetries are trivial if $\alpha = 2$.

2.4 In terms of the canonical coordinates $(r, s) = (xy, \frac{1}{2}x^2)$, the ODE reduces to

$$\frac{ds}{dr} = \frac{r}{1+r}.$$

2.5 Use canonical coordinates, for example,

$$(r, s) = \left(\frac{y}{x^3}, \ln|x|\right).$$

2.6 The generator is $X = \partial_x + y\partial_y$.

2.7 The generator is $X = y\partial_x$.

2.9 Use the linearized symmetry condition in the form (2.58) to prove this result.

Chapter 3

3.1 Start with $Q = \eta(x, y) - y'\xi(x, y)$; take the total derivative repeatedly to obtain $\eta^{(1)}, \eta^{(2)}, \ldots$.

3.2
$$\begin{aligned}\eta^{(4)} &= \eta_{xxxx} + (4\eta_{xxxy} - \xi_{xxxx})y' + (6\eta_{xxyy} - 4\xi_{xxxy})(y')^2 \\ &\quad + (4\eta_{xyyy} - 6\xi_{xxyy})(y')^3 + (\eta_{yyyy} - 4\xi_{xyyy})(y')^4 - \xi_{yyyy}(y')^5 \\ &\quad + \{6\eta_{xxy} - 4\xi_{xxx} + (12\eta_{xyy} - 18\xi_{xxy})y' + (6\eta_{yyy} - 24\xi_{xyy})(y')^2 \\ &\quad - 10\xi_{yyy}(y')^3\}y'' + \{3\eta_{yy} - 12\xi_{xy} - 15\xi_{yy}y'\}(y'')^2 \\ &\quad + \{4\eta_{xy} - 6\xi_{xx} + (4\eta_{yy} - 16\xi_{xy})y' - 10\xi_{yy}(y')^2 - 10\xi_y y''\}y''' \\ &\quad + \{\eta_y - 4\xi_x - 5\xi_y y'\}y^{(iv)}.\end{aligned}$$

3.3 (b)
$$\begin{aligned}X^{(4)} &= x\partial_x + \alpha y\partial_y + (\alpha-1)y'\partial_{y'} + (\alpha-2)y''\partial_{y''} \\ &\quad + (\alpha-3)y'''\partial_{y'''} + (\alpha-4)y^{(iv)}\partial_{y^{(iv)}}.\end{aligned}$$

(d)
$$\begin{aligned}X^{(4)} &= -y\partial_x + x\partial_y + (1+y'^2)\partial_{y'} + 3y'y''\partial_{y''} \\ &\quad + (4y'y''' + 3y''^2)\partial_{y'''} + (5y'y^{(iv)} + 10y''y''')\partial_{y^{(iv)}}.\end{aligned}$$

3.4 If $\alpha = 0$ then $\mathcal{L}$ is three-dimensional; otherwise, it is two-dimensional.

3.5 Every generator is of the form

$$X = c_1\partial_x + \left(c_2 e^x + c_3 e^{2x} + c_4 e^{-3x} + c_5 y\right)\partial_y.$$

3.6 Write down an equivalent ODE, using a new dependent variable $\tilde{y} = y - x$; find the symmetries of this ODE, then transform back to obtain the symmetries of the original ODE.

3.7 Use the linearized symmetry condition to show that

$$\xi = B(x), \qquad \eta = c(x)y + D(x),$$

and to derive conditions on $f(y)$. Solve these conditions for $c(x) = 0$ and $c(x) \neq 0$ separately.

Chapter 4

4.1 If $(r, s) = (y, x)$, the ODE reduces to the separable equation

$$\frac{dv}{dr} = -v^2 r^{-2},$$

where $v = \frac{ds}{dr}$. However, scaling symmetries reduce the original ODE to an apparently intractable first-order ODE.

4.2 The general solution is

$$y = \frac{c_1}{c_1^2(x + c_2)^2 - 1}$$

if $c_1 \neq 0$; the remaining solutions are

$$y = \frac{\pm 1}{2(x + c_2)}.$$

4.3 The generators are

$$X_1 = \partial_x, \qquad X_2 = \partial_y, \qquad X_3 = x\partial_x + \tfrac{1}{2}y\partial_y.$$

4.4 The Euler–Lagrange equation is

$$y'' = -\frac{y}{4x^2} - \frac{1}{\sqrt{x}y^2} - \frac{1}{y^3};$$

the scaling symmetries are generated by $X = x\partial_x + \frac{1}{2}y\partial_y$.

4.5 The characteristic is $Q = y^2 - xyy'$, so $Q = 0$ if and only if $y = cx$ for some $c \in \mathbb{R}$. The invariant solutions are $y = 0$ and $y = \pm x$.

4.6 The closed curve $x^2 + y^2 = 1$ is the only invariant solution; it partitions the set of solutions into two disjoint regions, as shown in Fig. 1.5.

Chapter 5

5.1 The most general such ODE is

$$y''' = x^{\alpha-3} F\left(x^{-\alpha}y,\ x^{1-\alpha}y',\ x^{2-\alpha}y''\right).$$

5.2 (a) $r = y/x$ and $v = x - y/y'$; (b) $r = xy$ and $v = x^2y'$. To find the most general third-order ODE, solve (c) to obtain (r_2, v_2), then calculate dv_2/dr_2 and apply the results of §5.1.

5.3 The ODE is

$$y'' = \frac{2(xy' - y)(1 + y'^2) + c(1 + y'^2)^{3/2}}{1 + x^2 + y^2}.$$

5.5 The Jacobi identity is not satisfied.

5.6 Span(X_1, X_2) is a Lie algebra if $\alpha = -1$. $\mathcal{L}(1)$ is three-dimensional and has a two-dimensional solvable Lie algebra. Otherwise, $\mathcal{L}(\alpha)$ is four-dimensional and has a three-dimensional solvable subalgebra. The $\mathfrak{sl}(2)$ subalgebra has a basis

$$X_a = y\partial_x, \qquad X_b = \tfrac{1}{2}x\partial_x - \tfrac{1}{2}y\partial_y, \qquad X_c = -x\partial_y.$$

5.7 The lowest-order common differential invariant is

$$I = \frac{2xy'' + y'}{y'^3}.$$

Chapter 6

6.1 The general solution is

$$x = y - c_1 \ln|y + c_1| + c_2;$$

there are also solutions of the form $y = c_1$.

6.2 The ODE has Lie symmetries generated by

$$X_1 = x^2\partial_x + xy\partial_y, \qquad X_2 = x\partial_x + 2y\partial_y.$$

Reduce the order using $(r_1, v_1) = (y/x,\ xy' - y)$.

6.5 Let $(r_1, v_1) = (x, y'/y)$; these are fundamental differential invariants for the group generated by X_1. If another set of differential invariants is used, it may appear necessary to write the solution parametrically.

6.6 The general solution is

$$\operatorname{erf}(y/\sqrt{2}) = c_1 + c_2/(x + c_3);$$

there are also solutions of the form

$$\operatorname{erf}(y/\sqrt{2}) = c_1 + c_2 x.$$

Chapter 7

7.1 Three functionally independent first integrals are

$$\phi^1 = y^{-2}y'' - 2y^{-3}y'^2, \qquad \phi^2 = 2y^{-3}y'' - 3y^{-4}y'^2,$$
$$\phi^3 = y^{-2}(xy'' - y') - 2xy^{-3}y'^2.$$

The general solution is

$$y = \left(c_1x^2 + c_2x + c_3\right)^{-1}.$$

7.2 The most general characteristic is

$$Q = \left\{(c_1x + c_2)e^y + (c_3x + c_4)e^{-y}\right\}y'^{1/2}$$
$$+\left\{c_5x^2 + c_6x + c_7\right\}y' + c_8x^2 + c_9x + c_{10}.$$

7.4 The groups generated by Q_8, Q_9 and Q_{10} are (respectively)

$$(\hat{x}, \hat{y}, \hat{y}') = \left(x + 2\varepsilon y',\ y + \varepsilon y'^2,\ y'\right),$$
$$(\hat{x}, \hat{y}, \hat{y}') = \left(x + 2\varepsilon\rho(xy' - y),\ y + \varepsilon\rho^2 y'^2(x - \varepsilon y),\ \rho y'\right),$$
$$(\hat{x}, \hat{y}, \hat{y}') = \left(\kappa x,\ \kappa^2 y - \varepsilon\kappa^2(2y - xy')^2,\ \kappa y'\right),$$

where

$$\rho = \frac{1}{1 - \varepsilon y'}, \qquad \kappa = \frac{1}{1 - \varepsilon(4y - 2xy')}.$$

7.6 If (7.86) holds then $J_j^{n-1} = 0$. The identity

$$\bar{D}J_j^k = -J_{j-1}^k - J_j^{k-1} - \omega_{y^{(k)}}J_j^{n-1} + \omega_{y^{(j)}}J_k^{n-1}$$

yields $J_j^k = 0,\ \forall\, j < k$ (by induction). The remaining integrability conditions follow from this result.

7.7 The characteristics are $Q_1 = 1$ and $Q_2 = y'$. There is only one cocharacteristic that is independent of y'', namely $\Lambda = e^x$.

Chapter 8

8.2 The Lie algebra is five-dimensional, because the general solution of the linearized symmetry condition is

$$\eta = c_1u + c_2, \qquad \xi = c_3x + c_4, \qquad \tau = (3c_3 - 2c_1)t + c_5.$$

8.3 The Lie point symmetry generators for the heat equation are

$$X_1 = \partial_x, \qquad X_2 = \partial_t, \qquad X_3 = u\partial_u, \qquad X_4 = x\partial_x + 2t\partial_t,$$
$$X_5 = 2t\partial_x - xu\partial_u, \qquad X_6 = 4xt\partial_x + 4t^2\partial_t - (x^2 + 2t)u\partial_u,$$
$$\left\{X_U = U(x,t)\partial_u : \; U_t = U_{xx}\right\}.$$

8.6 The Lie point symmetry generators are

$$X_1 = \partial_x, \qquad X_2 = \partial_t, \qquad X_3 = x\partial_x + t\partial_t, \qquad X_4 = t\partial_x + \partial_u,$$
$$\left\{X_{ZT} = Z(u,v)\partial_x + T(u,v)\partial_t : \; Z_u - uT_u + vT_v = 0,\right.$$
$$\left. Z_v - uT_v + T_u = 0\right\}.$$

A hodograph transformation (by which the dependent and independent variables are exchanged) linearizes the system.

Chapter 9

9.1 The PDE reduces to $v = F(r)$ where $v = t^k u$, $r = \sqrt{xt}$, and

$$F'' + \frac{1-2k}{r}F' - 4F = 0.$$

The general solution of this ODE is

$$F = r^k\left(c_1 I_k(2r) + c_2 K_k(2r)\right),$$

where $I_k(z)$, $K_k(z)$ are modified Bessel functions. When $k = \frac{1}{2}$, the solution is easier to write as

$$F = c_1 e^{2r} + c_2 e^{-2r}.$$

Rewriting the solution in the original variables, we obtain

$$u = \left(\frac{x}{t}\right)^{k/2}\left(c_1 I_k(2\sqrt{xt}) + c_2 K_k(2\sqrt{xt})\right).$$

These solutions can be extended to a large family with the aid of the remaining Lie symmetries. To obtain solutions of the Thomas equation, simply invert the linearizing transformation.

9.2 Use (8.25) to derive this identity. For systems, the corresponding result is

$$XQ_\alpha = Q_\beta \frac{\partial Q_\alpha}{\partial u_\beta}.$$

9.3 The travelling wave solutions are

$$u = c_1 + \frac{1}{c}\sin^{-1}\left(c_2 \exp\{-c(x - ct)\}\right).$$

9.4 The two-parameter family of solutions is

$$u = \frac{x + c_1}{t(\ln t + c_2)}.$$

Chapter 10

10.2 The optimal system consists of just one generator; for example, X_1 is optimal. The only invariant is

$$I = (\kappa^1)^2 + (\kappa^2)^2 + (\kappa^3)^2.$$

10.3 The invariants are

$$I^1 = (\kappa^4)^2 - 4\kappa^2\kappa^6,$$

which arises from an $\mathfrak{sl}(2)$ subalgebra, and

$$I^2 = \left((\kappa^4)^2 - 4\kappa^2\kappa^6\right)\left(\kappa^3 + \tfrac{1}{2}\kappa^4\right) + \kappa^1\kappa^4\kappa^5 - \kappa^2(\kappa^5)^2 - (\kappa^1)^2\kappa^6.$$

An optimal system of generators is

$$X_1, \qquad X_2 + kX_3, \qquad X_2 + kX_3 + X_6,$$
$$X_2 + X_5, \qquad X_2 - X_5, \qquad X_3, \qquad X_4 + kX_3,$$

where k is an arbitrary constant.

10.6 An optimal system of generators is

$$X_1, \qquad X_2 + \kappa^1 X_1 + \mu X_4, \qquad X_3 + \mu X_1, \qquad X_4 + \mu X_1, \qquad X_5,$$

where $\mu \in \{-1, 0, 1\}$ and κ^1 is an arbitrary constant.

Chapter 11

11.1 A set of inequivalent real-valued discrete symmetries is

$$(\hat{x}, \hat{y}) \in \{(x, y), (-x, -y), (1/x, y/x^2), (-1/x, -y/x^2)\}.$$

11.5 Every matrix B is equivalent to the identity matrix.

11.7 There are four real-valued inequivalent discrete symmetries of Burgers' equation:

$$(\hat{x}, \hat{t}, \hat{u}) \in \left((\alpha x, t, \alpha u), \left(\frac{\alpha x}{2t}, \frac{-1}{4t}, 2\alpha(ut - x) \right) \right), \qquad \alpha \in \{-1, 1\}.$$

11.10 The ODE has four inequivalent real contact symmetries, two of which are the point symmetries found in Example 11.4. The other two are

$$(\hat{x}, \hat{y}) \in \{(y', xy' - y), (-y', xy' - y)\}.$$

Bibliography

Anco, S. C. & Bluman, G. (1998). Integrating factors and first integrals of ordinary differential equations. *Eur. J. Appl. Math.*, **9**, 245–259.

Barenblatt, G. I. (1996). *Scaling, Self-similarity, and Intermediate Asymptotics.* New York: Cambridge University Press.

Bluman, G. W. & Kumei, S. (1989). *Symmetries and Differential Equations.* New York: Springer-Verlag.

Clarkson, P. A. (1996). Nonclassical symmetry reductions for the Boussinesq equation. *Chaos, Sol. Fractals*, **5**, 2261–2301.

Clarkson, P. A. & Olver, P. J. (1996). Symmetry and the Chazy equation. *J. Diff. Eqns.*, **124**, 225–246.

Cox, D. A., Little, J. B. & O'Shea, D. (1992). *Ideals, Varieties, and Algorithms.* New York: Springer-Verlag.

Fuchs, J. & Schweigert, C. (1997). *Symmetries, Lie Algebras and Representations.* New York: Cambridge University Press.

Golubitsky, M., Stewart, I. & Schaeffer, D. G. (1988). *Singularities and Groups in Bifurcation Theory, Vol. II.* New York: Springer-Verlag.

Hereman, W. (1996). Symbolic software for Lie symmetry analysis. In *CRC Handbook of Lie Group Analysis of Differential Equations. Vol. 3: New Trends in Theoretical Developments and Computational Methods*, ed. N. H. Ibragimov, pp. 367–413. Boca Raton: CRC Press.

Hydon, P. E. (1998a). Discrete point symmetries of ordinary differential equations. *Proc. Roy. Soc. Lond. A*, **454**, 1961–1972.

Hydon, P. E. (1998b). How to find discrete contact symmetries. *J. Nonlinear Math. Phys.*, **5**, 405–416.

Ibragimov, N. H. (ed.) (1994). *CRC Handbook of Lie Group Analysis of Differential Equations. Vol. 1: Symmetries, Exact Solutions, and Conservation Laws.* Boca Raton: CRC Press.

Ibragimov, N. H. (ed.) (1995). *CRC Handbook of Lie Group Analysis of Differential Equations. Vol. 2: Applications in Engineering and Physical Sciences.* Boca Raton: CRC Press.

Mansfield, E. L. & Clarkson, P. A. (1997). Applications of the differential algebra package `diffgrob2` to classical symmetries of differential equations. *J. Symb. Comp.*, **23**, 517–533.

Olver, P. J. (1993). *Applications of Lie Groups to Differential Equations.* 2nd ed. New York: Springer-Verlag.

Olver, P. J. (1995). *Equivalence, Invariants and Symmetry.* New York: Cambridge University Press.
Ovsiannikov, L. V. (1982). *Group Analysis of Differential Equations.* New York: Academic Press.
Sattinger, D. H. & Weaver, O. L. (1986). *Lie Groups and Algebras with Applications to Physics, Geometry, and Mechanics.* New York: Springer-Verlag.
Sewell, M. J. & Roulstone, I. (1994). Families of lift and contact transformations. *Proc. Roy. Soc. Lond. A*, **447**, 493–512.
Stephani, H. (1989). *Differential Equations: Their Solution Using Symmetries.* New York: Cambridge University Press.

Index

Printed in the United States
By Bookmasters